ISBN 978-09817374-6-1
0-9817374-6-3

This book was printed in the USA.
Published by:
JonHod Publishing Company
dnnsilver@aol.com

Contents

Introduction

Hello, and welcome to the wonderful world of music. This music instruction book teaches the important elements needed to learn basic music. For the sake of clarity, it is broken up into four logical sections made up of specific lessons, followed by a review in the form of questions, music and listening exercises and puzzles. Each week a different lesson or parts of a lesson can be learned in one school year. If music instruction is only given one-half of a school year, then two lessons of sections I through III and parts of section IV can be learned per week.

The first section pertains to melody and accompaniment and how it relates to the keyboard. It introduces music notation in both the treble and bass clefs and consists of basic finger exercises that help students make the connection between reading music notation and playing the keyboard.

The second section focuses on rhythm. In addition to learning to play the keyboards, the students learn to maintain rhythms by clapping or tapping out rhythms on a table. First, they are given a brief history lesson about the drum then they are introduced to a seven-piece drum set. In addition to this, the students learn the difference between duple and triple meters and how to keep a beat in these two meters.

The third section asks students to take information learned in sections one and two, and create original compositions. This section focuses on creativity and how it is organized into intelligible music. The students take poetry they have created and set it to original music using melody, accompaniment and rhythm and or take music they have created and add lyrics. Once the compositions are completed, they can be recorded and performed. If there is no keyboard available, the students can hum or sing a melody, record it and then add words. This stimulates original ideas and logical thought patterns.

The last section is history which begins with suggested listening material followed by brief music history lessons in European and African American music. The information is set up in chronological order so that the students learn when different styles of music took place, how each style evolved and how it was affected by social changes during different time periods. These music history lessons bridge the gap between today's music and the music of yesterday.

The students learn how the music they enjoy today developed out of yesterday's music. This understanding helps them to develop a greater appreciation for various types of music and the people that created it.

This music instruction book is a thorough teaching guide for music students interested in learning the fundamentals of music as well as a guide that supplements material already used by budding musicians. It is equipped with a piano chart (located in the back of the book) to familiarize the students with a keyboard's make up and also an audio recording is provided to measure the students' musical growth. Because of these tools and the information provided, this guide is designed to serve as a foundation for a formal music education and diverse music experience for music students of all levels and ages.

About the Piano

The music instrument called the piano became popular during the nineteenth century. Musicians used the piano before the nineteenth century however, it did not have the ability to reach various dynamic levels until it had been developed to do so in the early 1800's.

The keyboard of the piano, from which the instrument is controlled, consists of eighty-eight keys (fifty-two white keys, thirty-six black keys). When the keys are pressed, felt-covered hammers strike the steel strings, which are located inside the piano. This produces the tones we hear.

The piano comes in various forms. The "grand" piano, which is shaped like a horizontal harp, is most commonly used for recitals or concerts. The "spinet" is a rectangular shaped piano and the "upright" piano is also rectangular in shape, but its case is taller than that of the spinet.

There are various uses for the piano. It is used as an accompanying instrument as well as a solo instrument. Many instrumentalists encounter their first musical experience on the piano; therefore, it has served as the primary instrument for learning other instruments.

No matter what the reason may be for becoming acquainted with the piano, mastering this timeless instrument can prove to be a rich and rewarding experience that lasts a lifetime.

Finger Chart

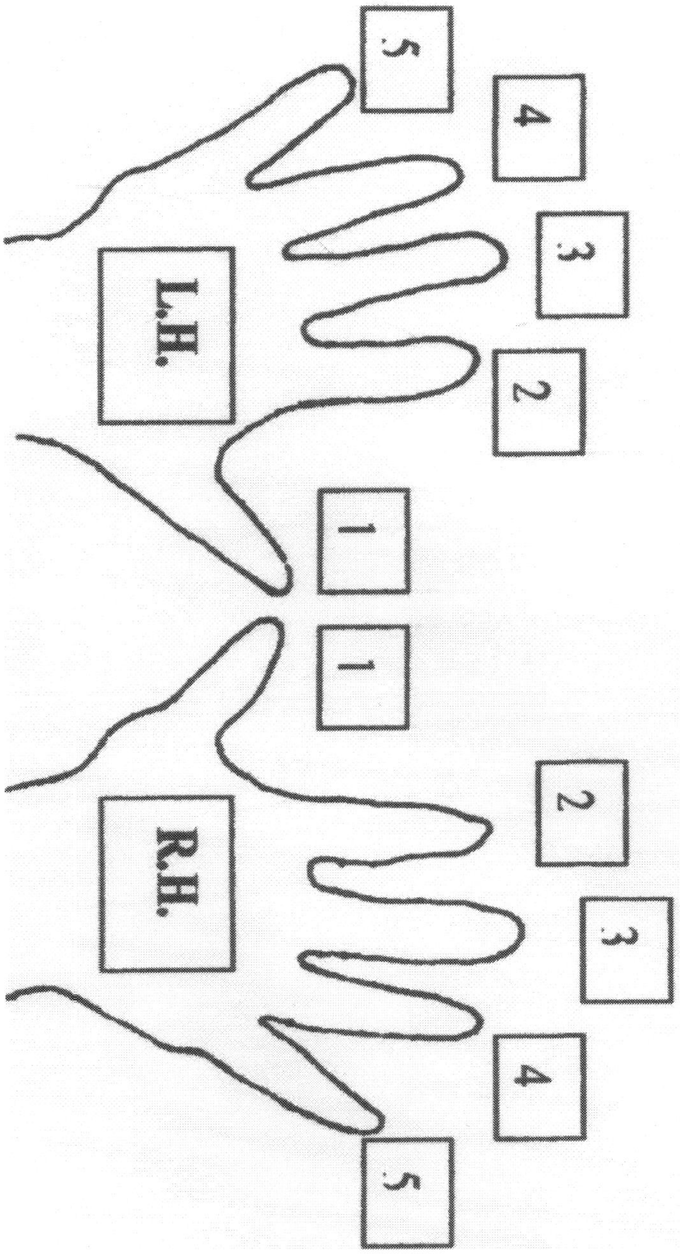

The Music Alphabet

The ***music alphabet*** consists of the first seven letters of the alphabet: A-G. Following the note, "G," is "A."

ABCDEFGABCDEFG

The Music Alphabet on the Piano

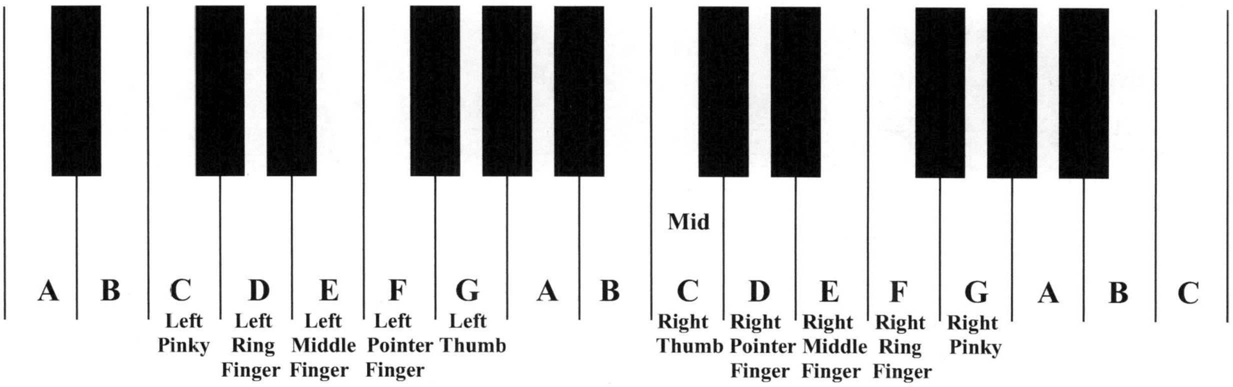

Lesson 2
Symbols and Terms

Staff: Five (5) lines and four (4) on which the clefs and notes are placed

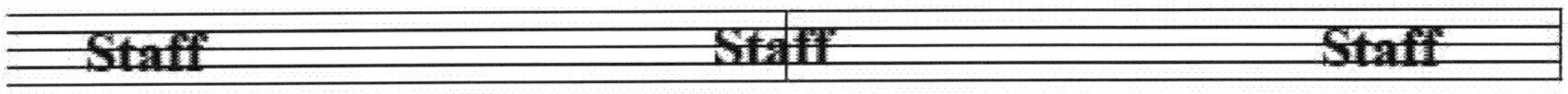

Measure: a unit of space on the staff, separated by barlines

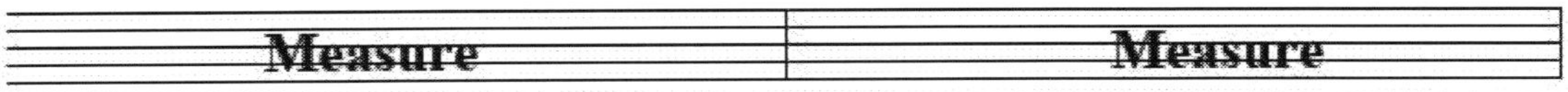

<u>Barline</u>: the symbol that separates measures

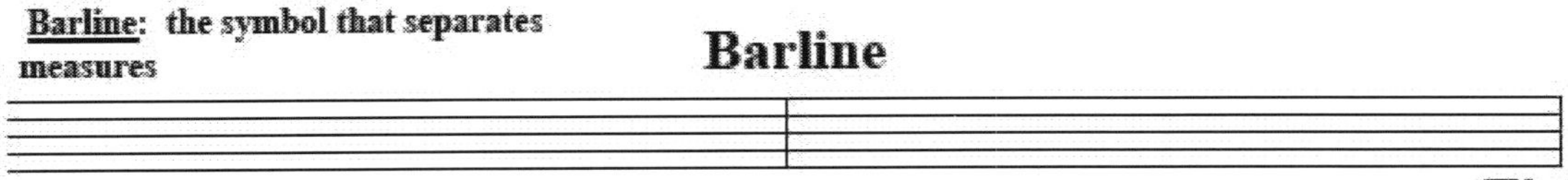

<u>Fine Barline</u>: the symbol placed at the end of a passage of music indicating the end of that passage

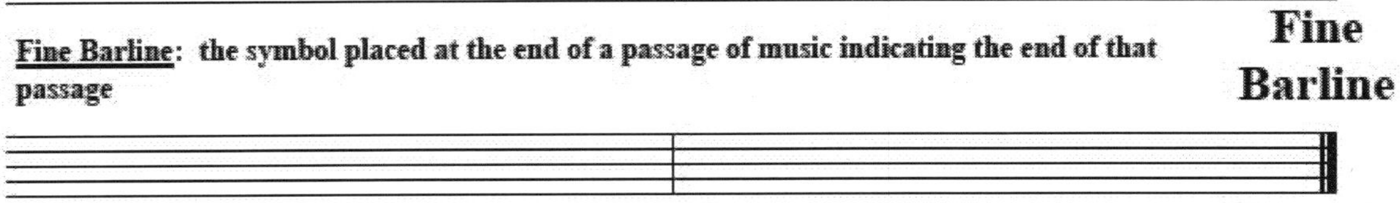

<u>Repeat Sign</u>: the symbol placed at the end of a song instructing to play or sing the passage of music again

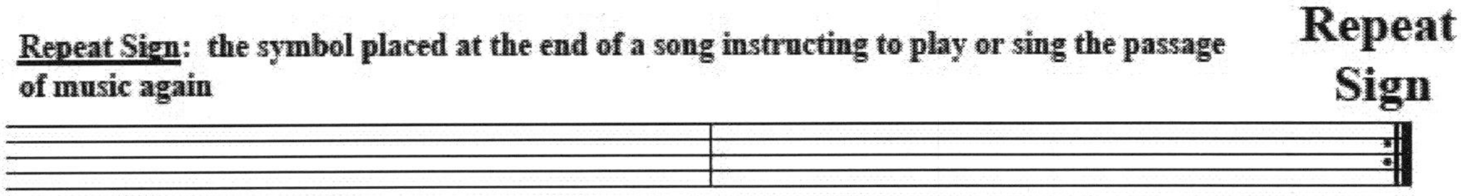

<u>Treble Clef</u>: The symbol placed on a staff that indicates the pitch of a note. The pitches of the notes placed on the treble clef staff range from middle to high.

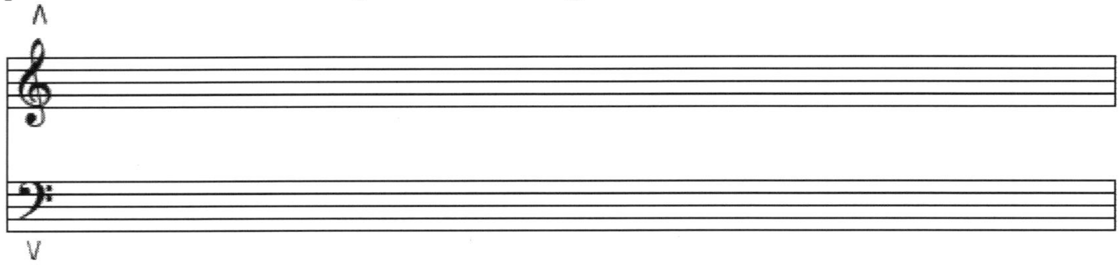

<u>Bass Clef</u>: The symbol placed on a staff that indicates the pitch of a note. The pitches of the notes placed on the treble clef staff range from middle to low.

Draw the Following Symbols

Staff:

Measure:

Bar line:

Fine Bar line:

Repeat Sign:

Treble Clef or "G" Clef:

Bass Clef or "F" Clef:

Lesson 3

Learning Notes and Note Values

Each note has its own value. Press any key on the piano or keyboard or sing any note and hold it for the value of each type of note.

Quarter Note = 1 beat

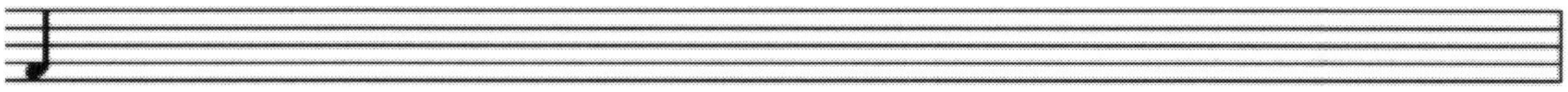

Half Note = 2 beats

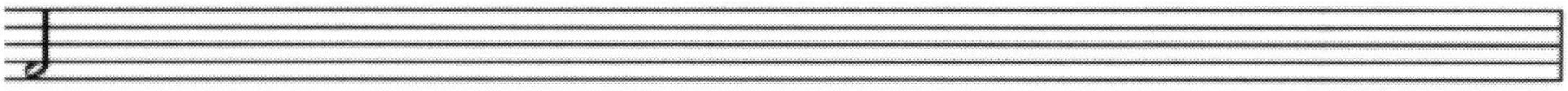

Dotted Half Note = 3 beats

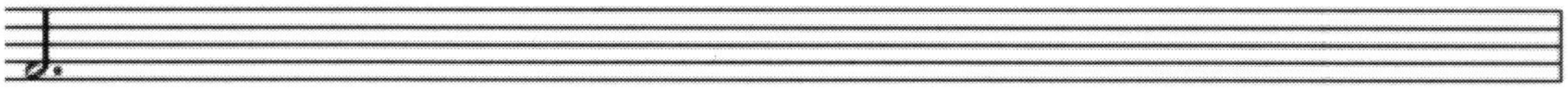

Whole Note = 4 beats

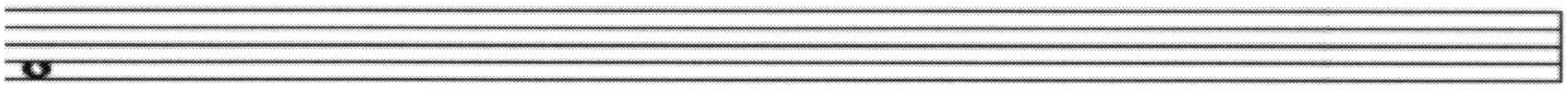

Adding Notes

Directions: Draw the notes listed below then add them together. Write the sum in the space provided.

Ex: Quarter Note + Quarter Note

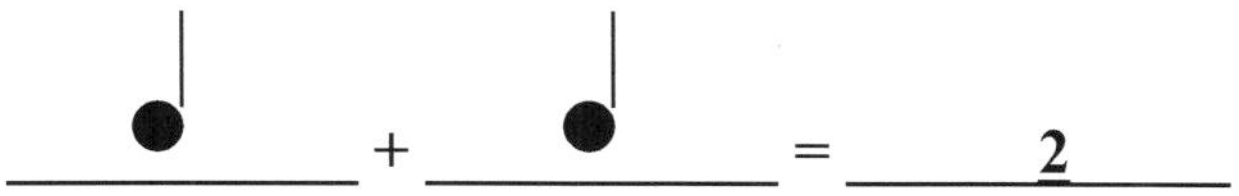

1. **Quarter Note + Half Note**		+		=
2. **Quarter Note + Whole Note**		+		=
3. **Half Note + Dotted Half Note**		+		=
4. **Dotted Half Note + Whole Note**		+		=
5. **Whole Note + Whole Note**		+		=
6. **Dotted Half Note + Dotted Half Note**		+		=
7. **Whole Note + Half Note**		+		=
8. **Quarter Note + Half Note + Whole Note**		+		=
9. **Quarter Note + Dotted Half Note + Quarter Note**		+		=
10. **Quarter Note + Dotted Half Note + Quarter Note**		+		=
11. **Dotted Half Note + Half Note + Whole Note**		+		=

Lesson 4

"C" position for right hand: finger number one (thumb) is placed on "Middle C", finger number two (pointer finger) is placed on "D", finger number three (middle finger) is placed on "E", finger number four (ring finger) is placed on "F" and finger number five (pinky) is placed on "G".

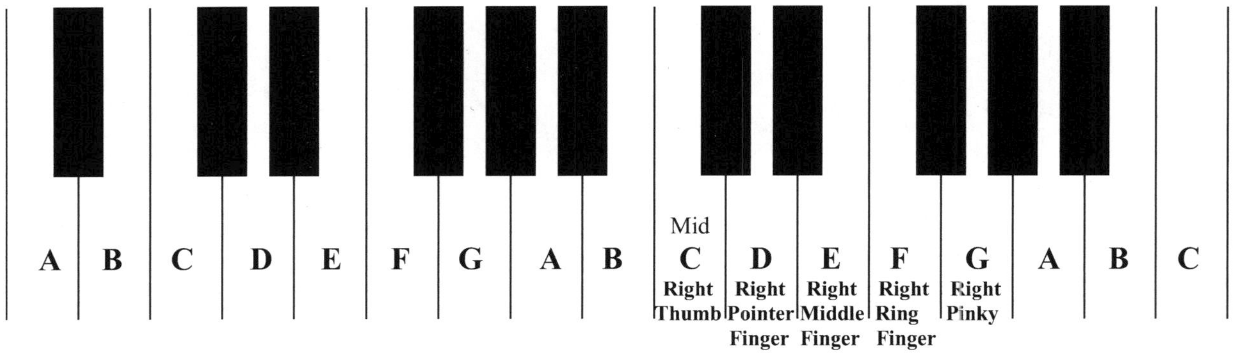

Learning the Treble Clef: Middle "C"-G

Directions: Write the letter name under each treble clef note. Once the letter names are written, place your fingers on the corresponding notes on the piano chart as if you are playing the notes on a piano.

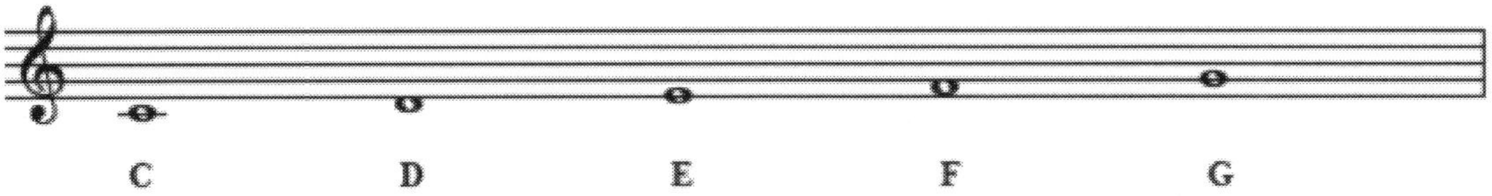

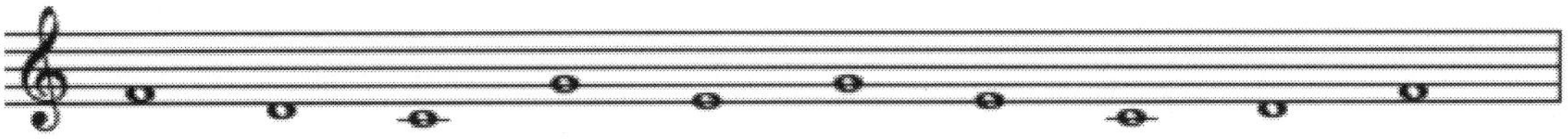

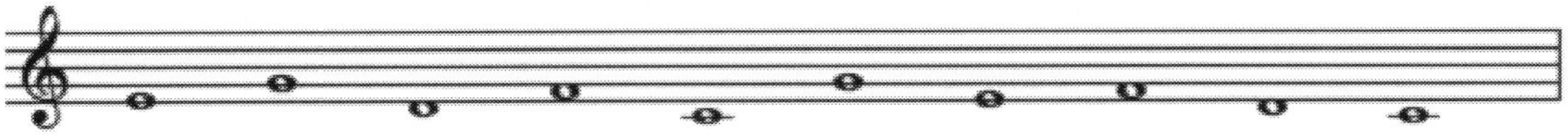

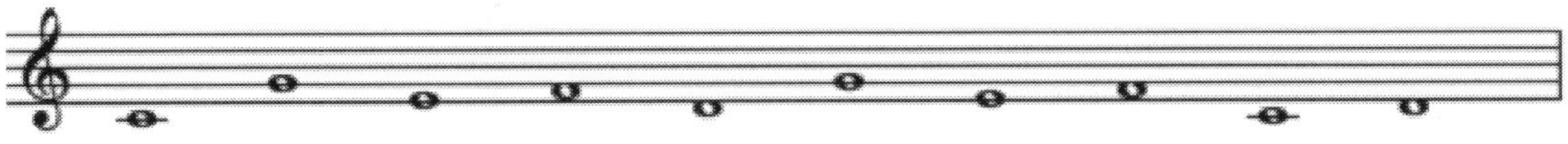

Learning the Treble Clef Middle "C"-G II

Directions: Recite the following notes in your head until you are able to read them without looking at the chart on the previous page. Recite them backwards, forwards and out of order until you recognize each note. Then place your hands in "C" position (as illustrated in this lesson) and play the notes. **Do not write the letter names of the notes.**

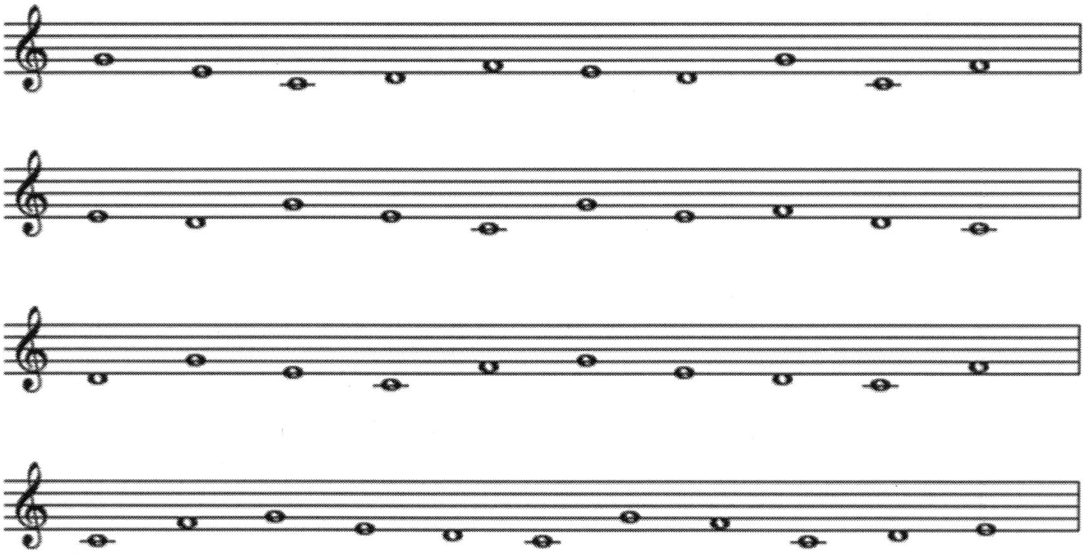

Treble Clef (Track 1)

Directions: Play the notes in exercises 1, 2, 3 and 4 on the piano or keyboard (if you don't have either, place your fingers on the corresponding notes on the piano chart as if you're playing the notes on a piano). Then once you're comfortable playing the notes, sing the notes using the letter names.

Exercise 1

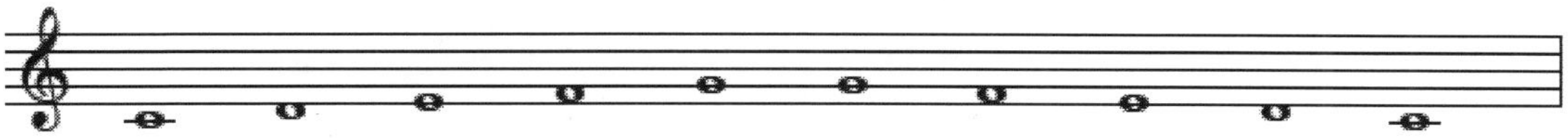

Exercise 2

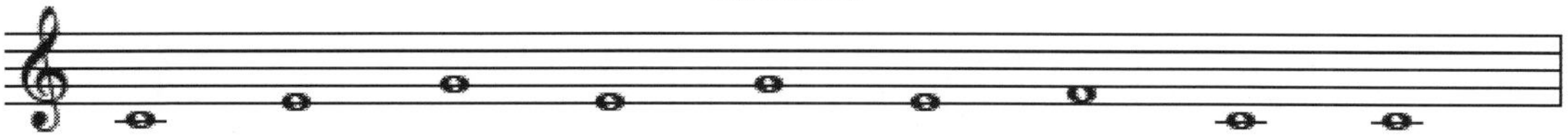

Exercise 3

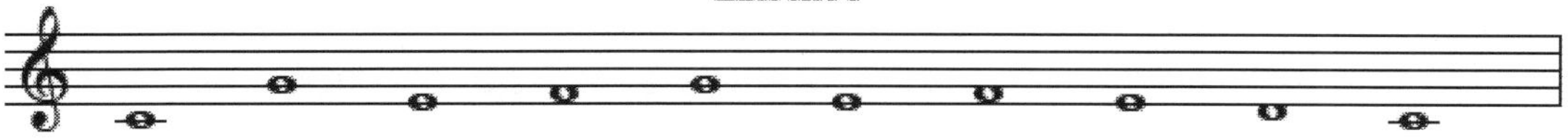

Exercise 4

Sing and Play Treble Clef (Track 2)

Directions: Play the notes of the scale below. (If you don't have a piano or keyboard, place your fingers on the corresponding notes on the piano chart as if you are playing a piano.) Then once you're comfortable playing the notes, sing the letter names of the notes while playing the notes. Make sure you hold the notes for their correct value.

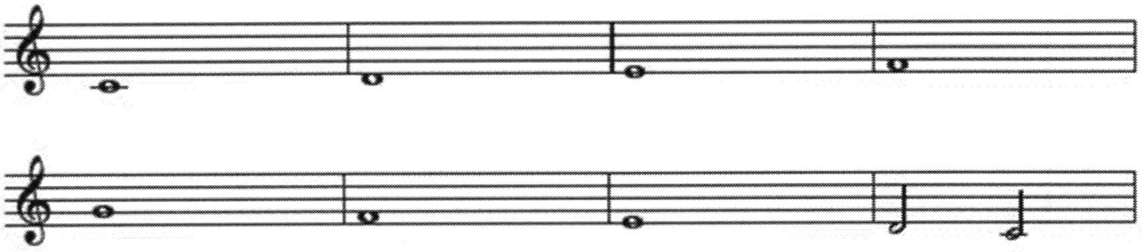

Directions: Once you've completed the exercises above, play songs #1 and #2 (holding each note for its correct value). Once the songs can be played comfortably, sing the songs using the letter names first, then using the syllable, "la." Once the songs can be sung and played comfortably, make up words to the songs using one syllable per note to create your own song.

Song # 1

Song # 2

Lesson 5

"C" position for left hand: finger number five (pinky) is placed on "C", finger number four (ring finger) is placed on "D", finger number three (middle finger) is placed on "E", finger number two (pointer finger) is placed on "F" and finger number one (thumb) is placed on "G".

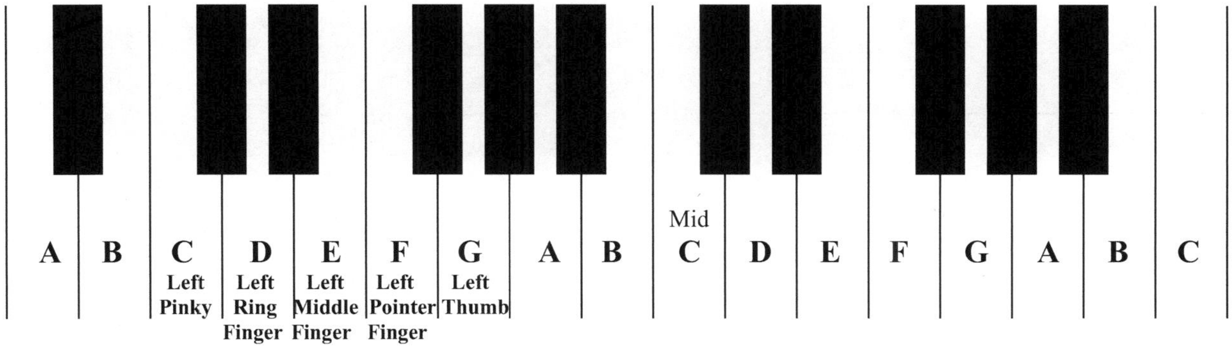

Learning the Bass Clef: "C"-G

Directions: Write the letter name under each treble clef note. Once the letter names are written, place your fingers on the corresponding notes on the piano chart as if you are playing the notes on a piano.

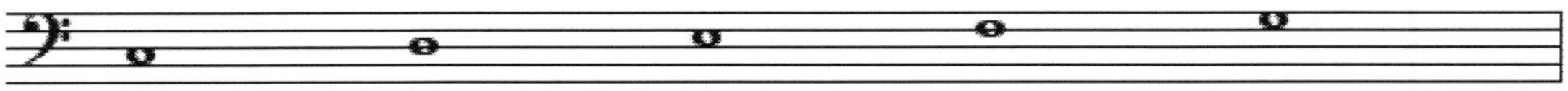

C D E F G

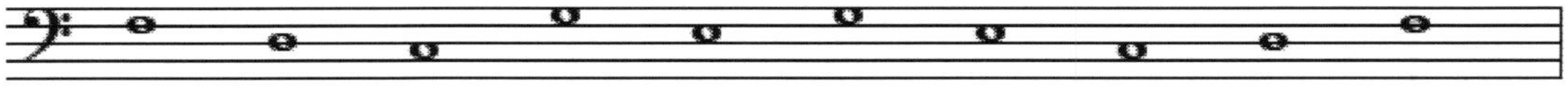

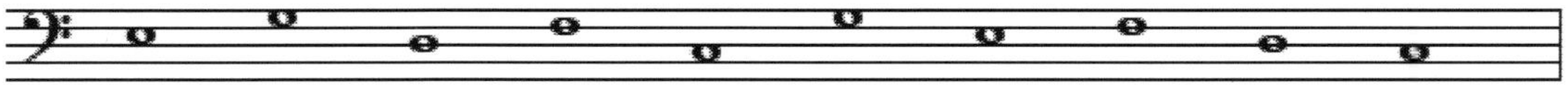

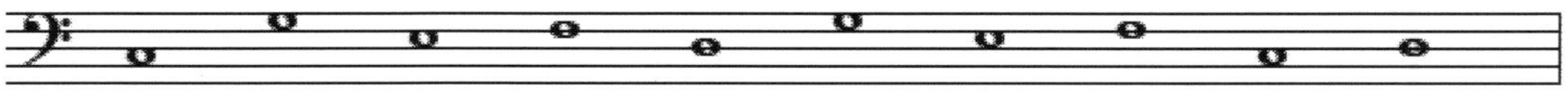

Learning the Bass Clef "C"-G II

Directions: Recite the following notes in your head until you are able to read them without looking at the chart on page on the previous page. Recite them backwards, forwards and out of order until you recognize each note. **Do not write the letter names of the notes**.

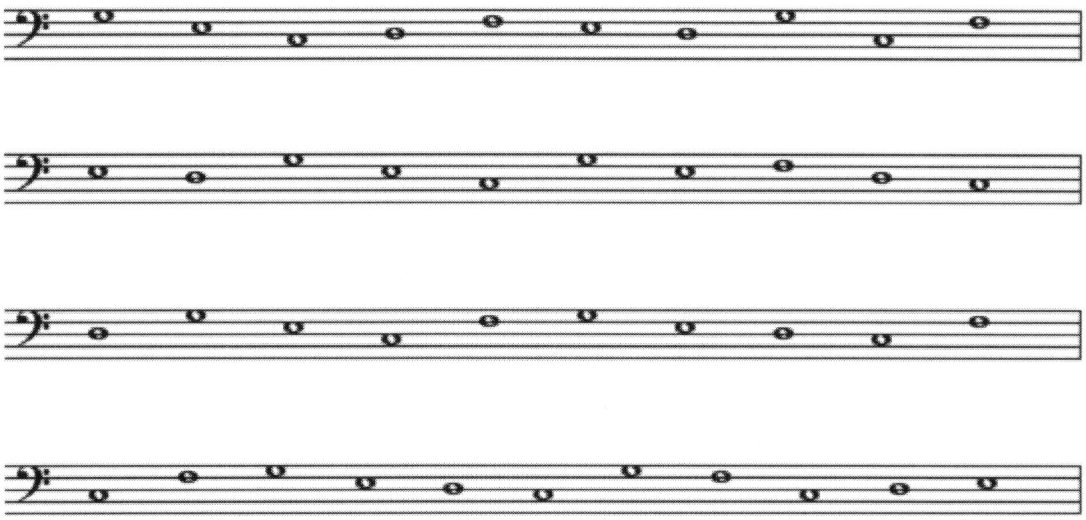

Bass Clef (Track 3)

Directions: Play the notes in exercises 1, 2, 3 and 4 on the piano or keyboard (if you don't have either, place your fingers on the corresponding notes on the piano chart as if you're playing the notes on a piano). Then once you're comfortable playing the notes, sing the notes using the letter names.

Note: Bass clef notes are lower in range than treble clef notes therefore, sing the notes in your own range.

Exercise 1

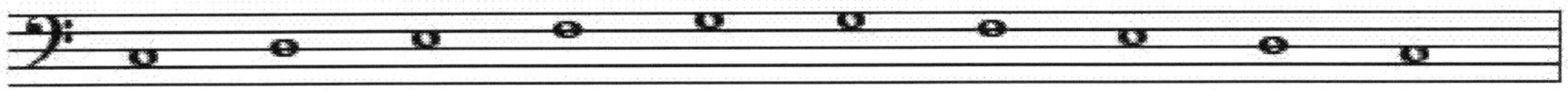

Exercise 2

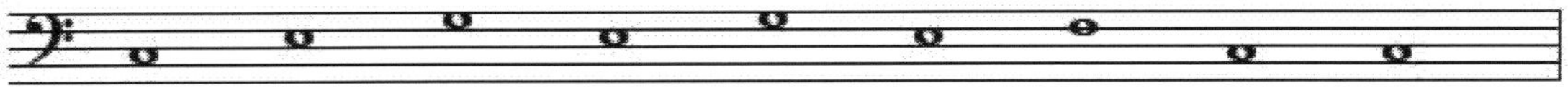

Exercise 3

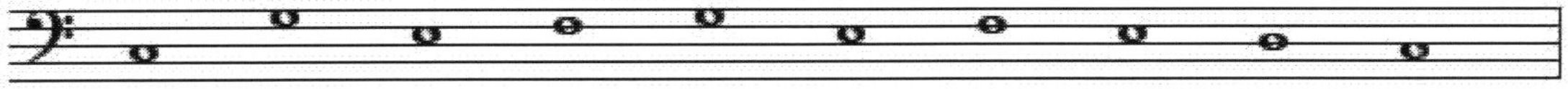

Exercise 4

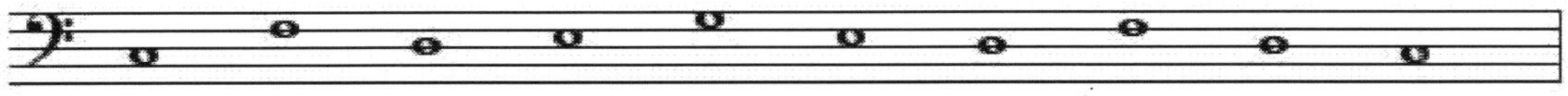

Sing and Play Bass Clef (Track 4)

Directions: Play the notes of the scale below. (If you don't have a piano or keyboard, place your fingers on the corresponding notes on the piano chart as if you are playing a piano.) Then once you're comfortable playing the notes, sing the letter names of the notes while playing the notes. Make sure you hold the notes for their correct value.

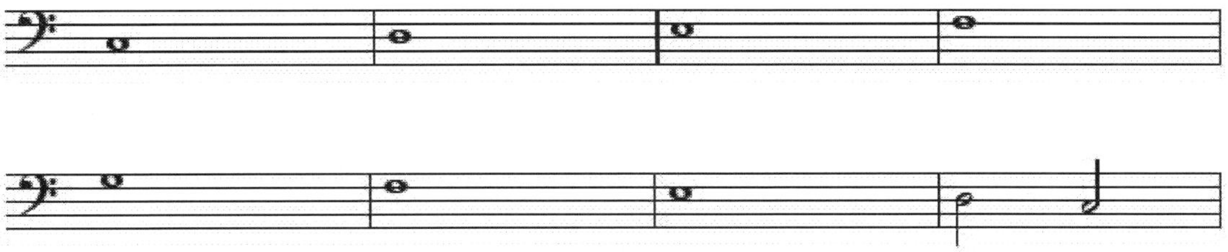

Directions: Once you've completed the exercises above, play songs #1 and #2 (holding each note for its correct value). Once the songs can be played comfortably, sing the songs using the letter names first, then using the syllable, "la." Once the songs can be sung and played comfortably, make up words to the songs using one syllable per note to create your own song.

Song #1

Song #2

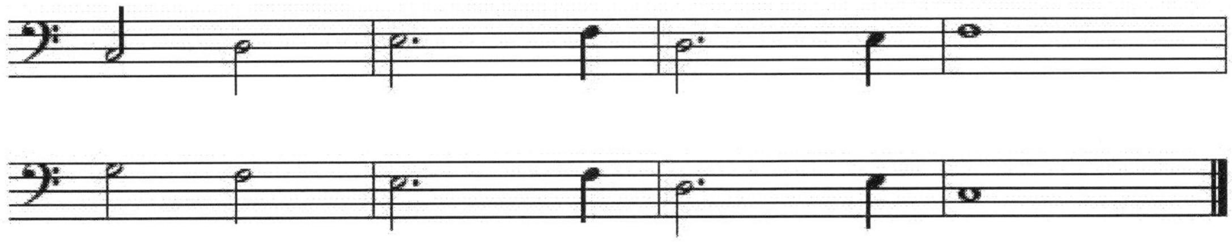

Lesson 6

Sing & Play Treble and Bass Clef (Track 5)

Sing & Play Treble & Bass Clef II (Track 6)

Lesson 7

Sing & Play Hands Together (Track 7)

Song #1

Song #2

Sing & Play Hands Together II (Track 8)

Song #1

Song #2

Sing and Play Using Eighth Notes Hands Together (Track 9)

Eighth Notes: One eighth note equals 1/2 of a quarter note. Therefore, two eighth notes equal one quarter note. Review Song #1 then play the Eighth Note Exercise in order to understand the 2-1 relationship between eighth notes and quarter notes. After playing the Eighth Note exercise, play Song #1 Using Eighth Notes and compare it to Song #1.

Song #1 Review

Eighth Note Exercise

This exercise shows two eighth notes to every quarter note.

Song #1 Using Eighth Notes

Sing and Play Hands Together Using Eighth Notes II (Track 10)

Song #1

Song #1 Using Eighth Notes

Lesson 8

Learning Solid and Broken Bass Clef Chords

Diagram of Bass Clef Chords

Note: The following diagrams illustrate the left hand position needed to play a "C" and "G7" chord. The notes are to be played simultaneously.

"C" Chord

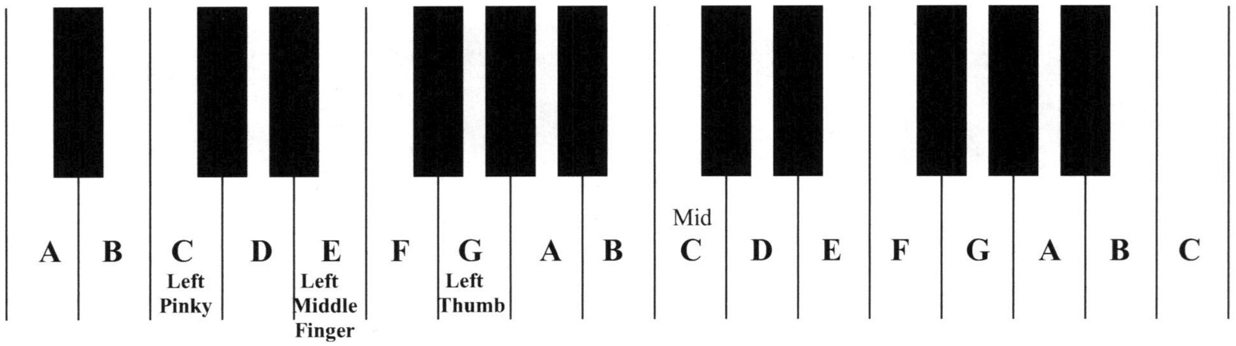

"G7" Chord

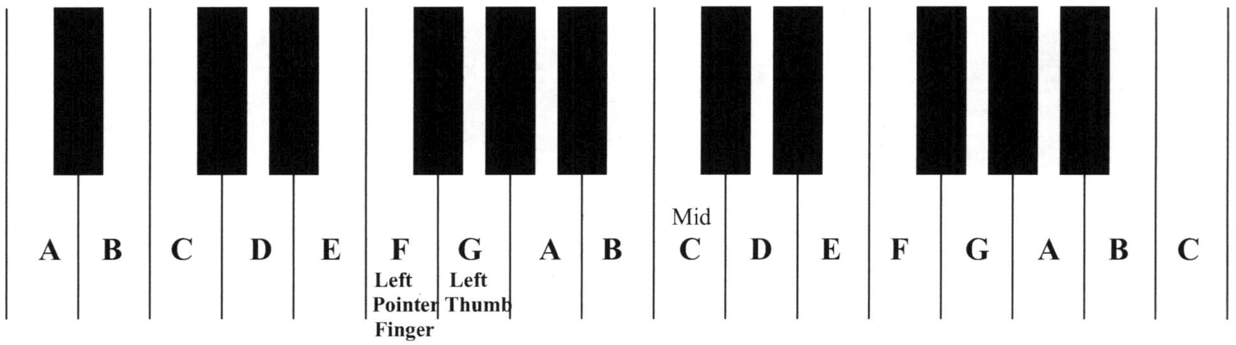

Learning Solid Bass Clef Chords (Track 11)

Solid Chord: A solid chord is made up of two or more notes played at the same time to create harmony in music. The following bass clef staves illustrate two solid chords: a "C" chord, which is made up of the notes C-E-G and a "G7" chord which is made up of the notes F-G. Place your left hand in "C" position on the piano keyboard or piano chart and follow the diagrams below in order to learn how to play the two chords. Play the chords shown until they are played comfortably. Be sure to hold them for their correct value.

"C" Chord

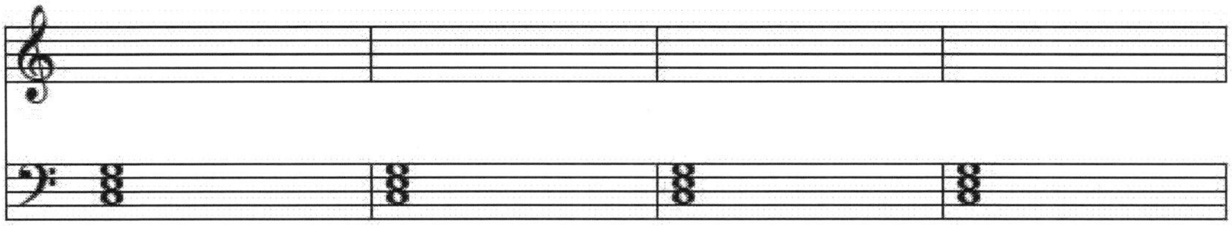

"G7" Chord

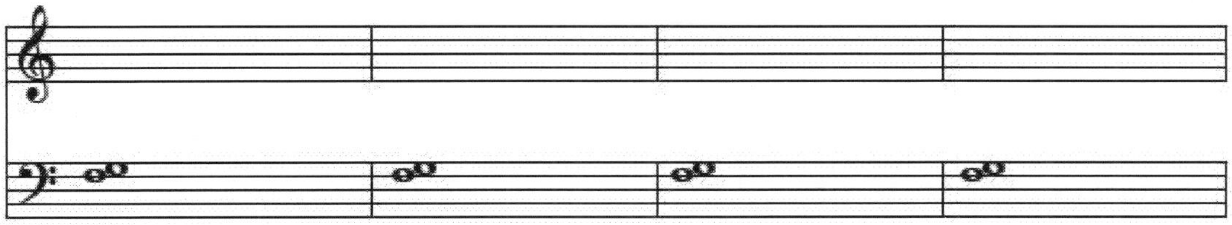

"C" Chord

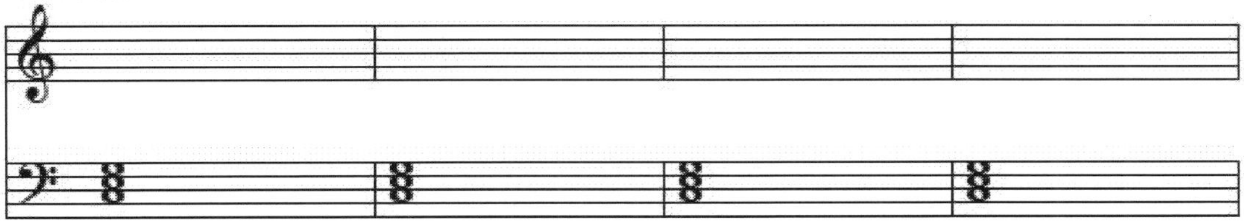

Sing and Play with Solid Bass Clef Chords (Track 12)

Song #1

Song #2

Learning Broken Bass Clef Chords (Track 13)

Broken Chord: A broken chord is made up of two or more notes played one after the other. The following bass clef staves illustrate two broken chords: a "C" chord, which is made up of the notes C-E-G and a "G7" chord which is made up of the notes F-G. Place your left hand in "C" position on the piano keyboard or piano chart and follow the diagrams below in order to learn how to play the two chords. Play the chords shown until they are played comfortably. Be sure to hold them for their correct value.

"C" Chord

"G7" Chord

"C" Chord

Sing and Play with Broken Bass Clef Chords (Track 14)

Song #1

Song #2

Lesson 9

Learning Solid Treble Clef Chords (Track 15)

Directions: Place your right hand in "C" position on the piano keyboard or piano chart and follow the diagrams below in order to learn how to play the two chords. Play the chords shown until they are played comfortably. Be sure to hold them for their correct value.

"C" Chord

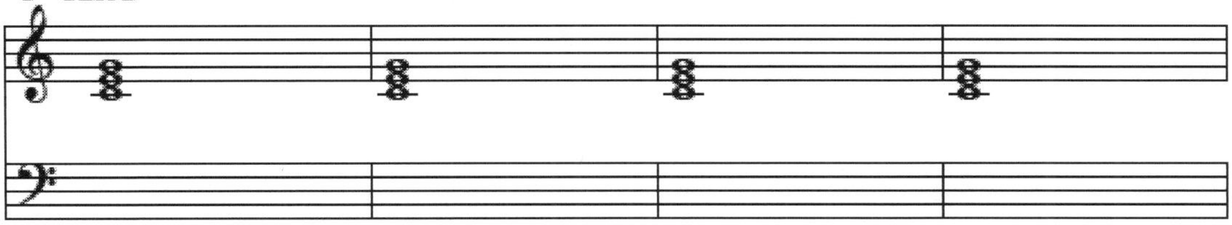

"G7" Chord

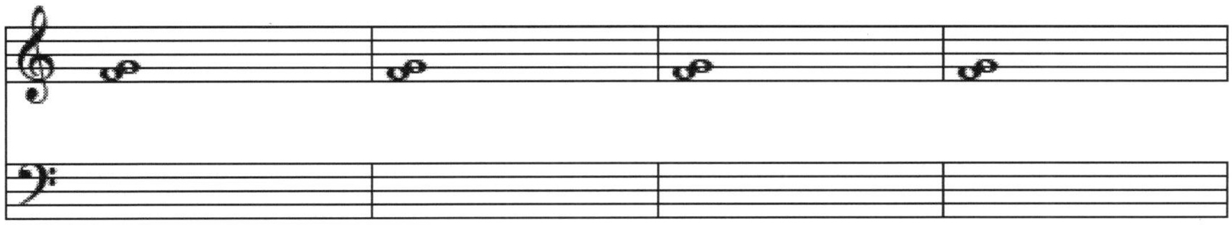

"C" Chord

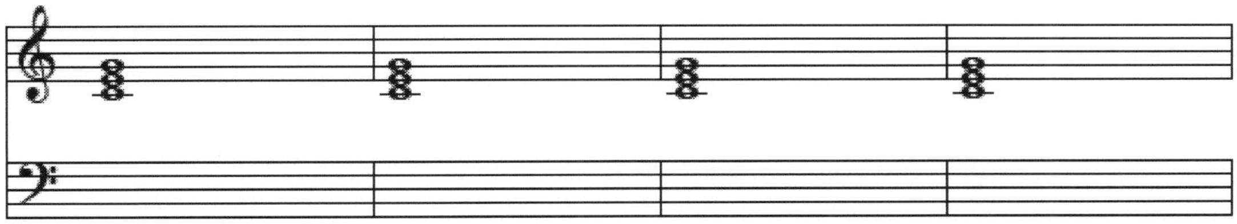

Sing and Play with Solid Treble Clef Chord (Track 16)

Song #1

Song #2 Using Eighth Notes

Learning Broken Treble Clef Chords (Track 17)

Directions: Place your right hand in "C" position on the piano keyboard or piano chart and follow the diagrams below in order to learn how to play the two chords. Play the chords shown until they are played comfortably. Be sure to hold them for their correct value.

"C" Chord

"G7" Chord

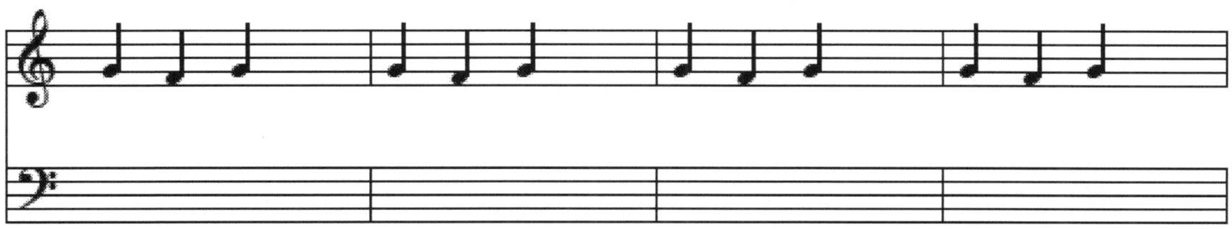

"C" Chord

Sing and Play with Broken Treble Clef Chords (Track 18)

Song #1

Song #2

Lesson 10

Learning Broken Treble Clef Chords with Solid Bass Clef Chords (Track 19)

Directions: Place your right hand in "C" position on the piano keyboard or piano chart and follow the diagrams below in order to learn how to play the two chords. Play the chords shown until they are played comfortably. Be sure to hold them for their correct value.

"C" Chord

"G7" Chord

"C" Chord

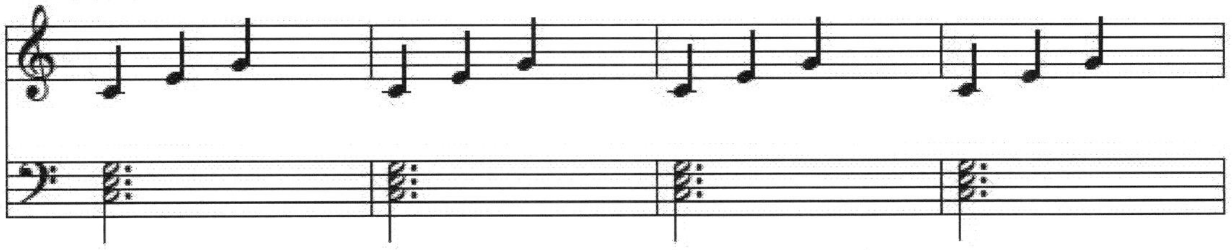

Learning Solid Treble Clef Chords with Broken Bass Clef Chords (Track 20)

Directions: Place your right hand in "C" position on the piano keyboard or piano chart and follow the diagrams below in order to learn how to play the two chords. Play the chords shown until they are played comfortably. Be sure to hold them for their correct value.

"C" Chord

"G7" Chord

"C" Chord

Sing and Play with Broken and Solid Treble and Bass Clef Chords (Track 21)

Lesson 11

Swaying (Track 22)

Repeat this line only.

Breezes (Track 23)

Lesson 12

Notes on the Treble Clef Beginning with Middle "C"

Note: Now that notes Middle "C" - G of the treble clef are learned, it's time to learn the remaining notes of the treble clef. The notes on the staff follow a pattern of line, space, line, space. As you scale up the staff, following the pattern, you say the alphabet in ascending order. As you scale down the staff, following the pattern, you say the alphabet in descending order.

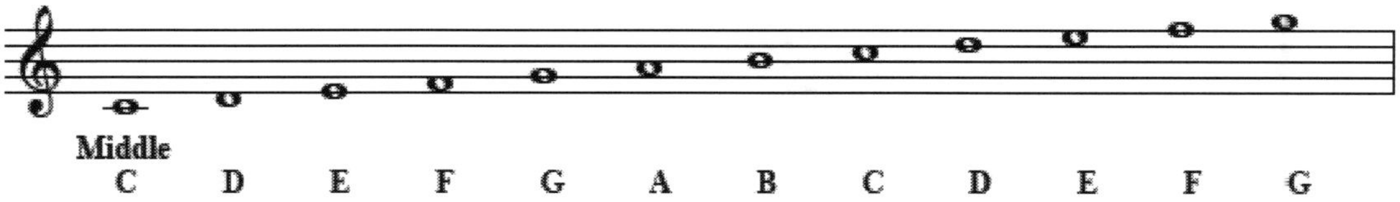

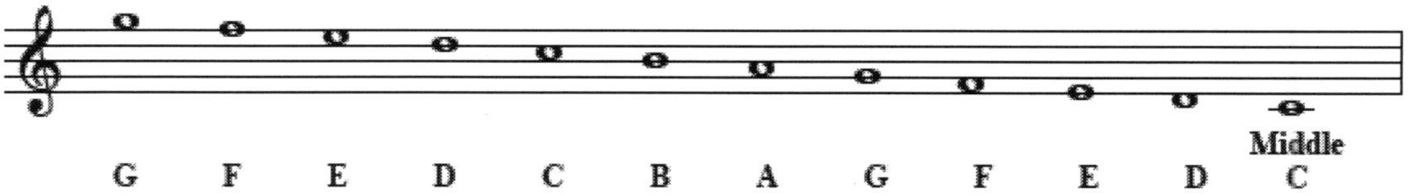

Treble Clef Exercise

Directions: Figure out the names of the notes by scaling up and down the staff. Write the name of the note under each treble clef note. **Try not to use the chart on the previous page.**

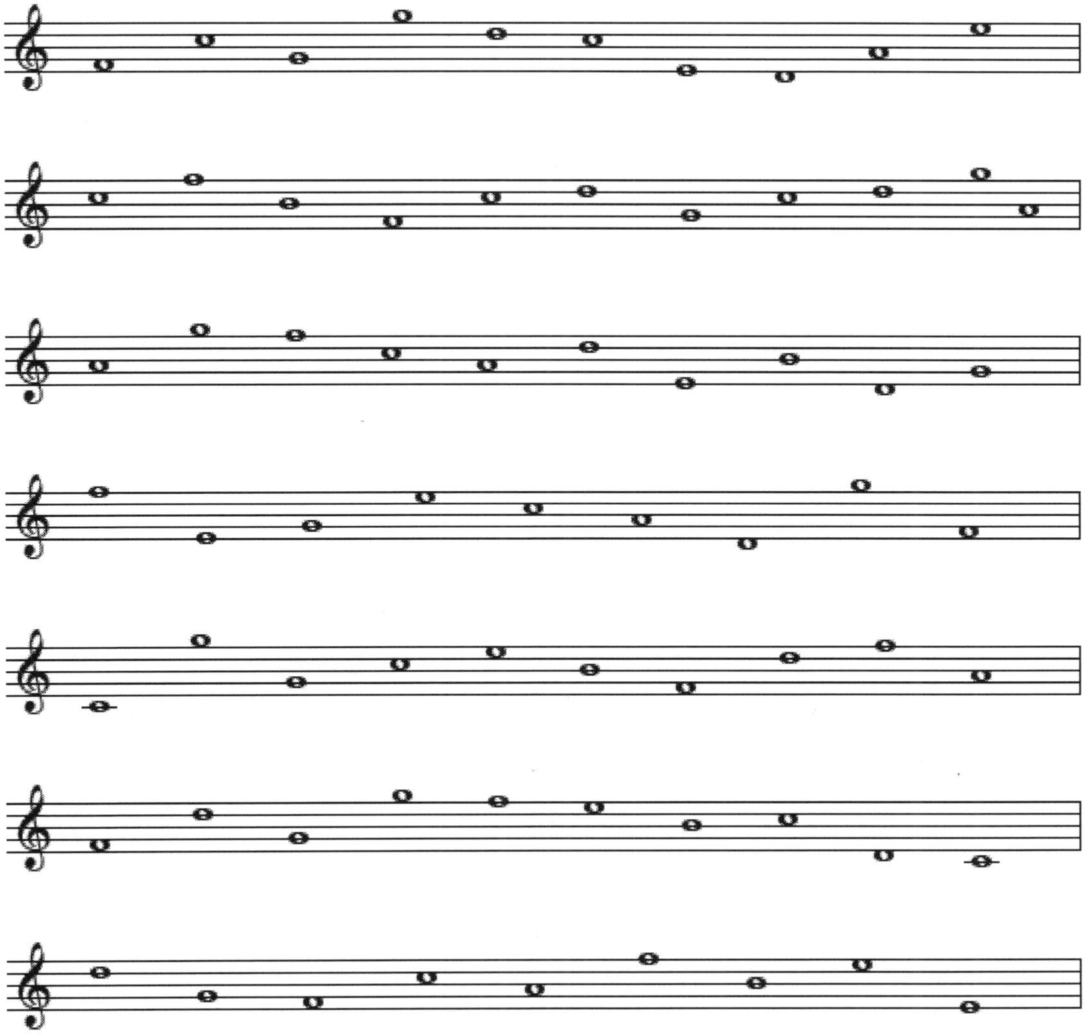

Treble Clef Exercise II

Directions: Draw the notes of the corresponding letter name on the staff. **Try not to use the chart on the previous page.**

G D C A E B F Middle C G F

Middle C F D B C E G D E G

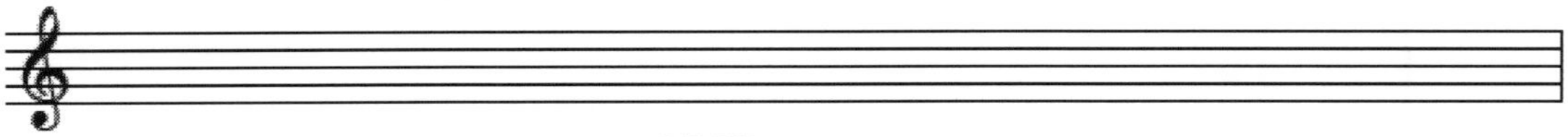

F C B D Middle C G F A G E

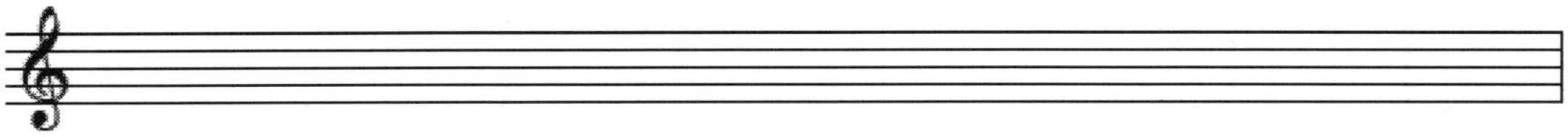

A D E F B G E C D F

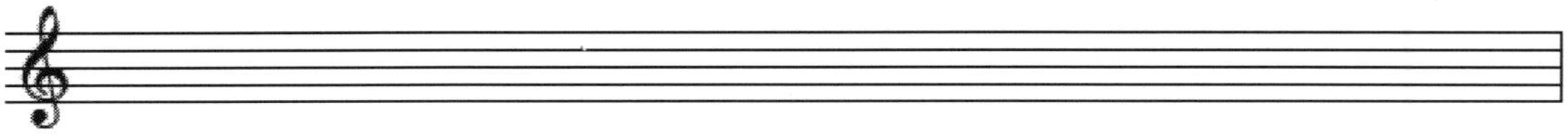

D C B E C B C D A G

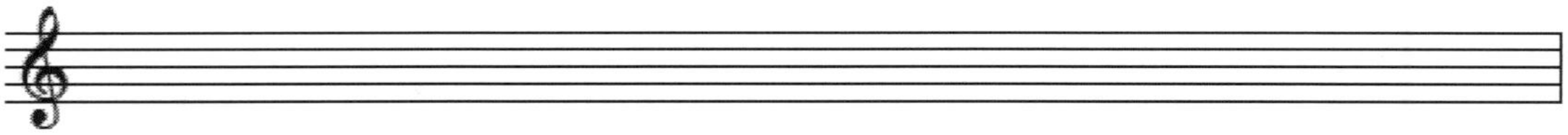

B D F E G A F A B Middle C

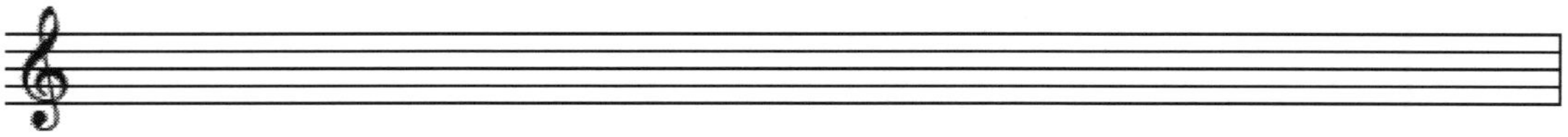

G C A D E G B F G E

Lesson 13

Notes in the Bass Clef Ending with Middle "C"

Note: Now that notes "C" - "G" of the bass clef are learned, it's time to learn the remaining notes of the bass clef. The notes on the staff follow a pattern of line, space, line, space (just like the notes on the treble staff). As you scale up the staff, following the pattern, you say the alphabet in ascending order. As you scale down the staff, following the pattern, you say the alphabet in descending order.

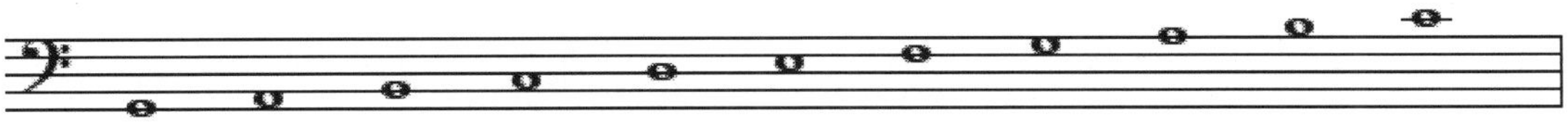

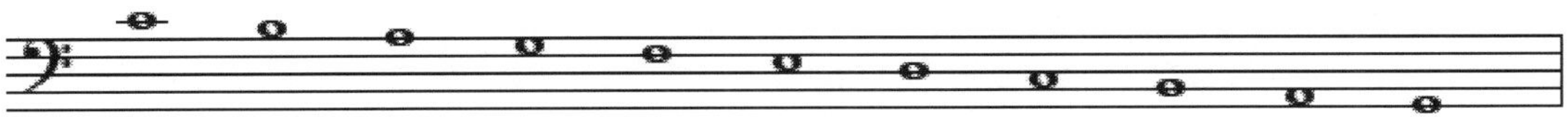

Bass Clef Exercise

Directions: Figure out the names of the notes by scaling up and down the staff. Write the name of the note under each treble clef note. **Try not to use the chart on the previous page.**

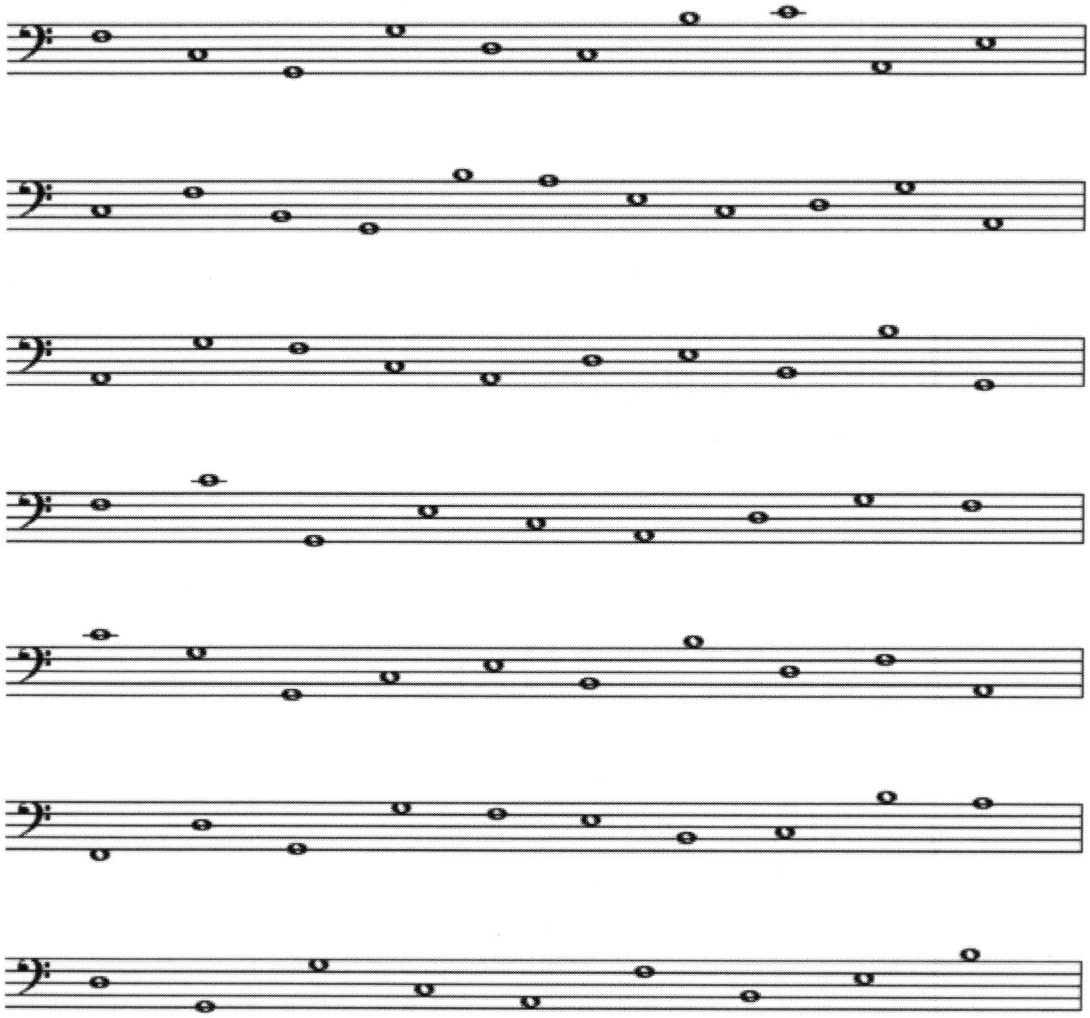

Bass Clef Exercise II

Directions: Draw the notes of the corresponding letter name on the staff. **Try not to use the chart on the previous page.**

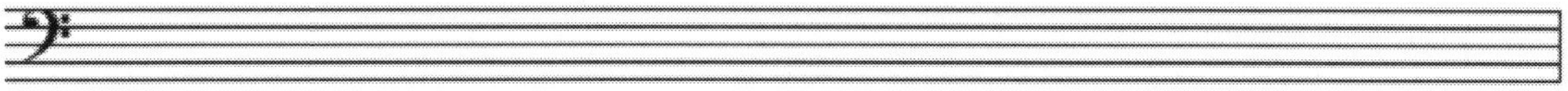

G D C A E B F Middle C G F

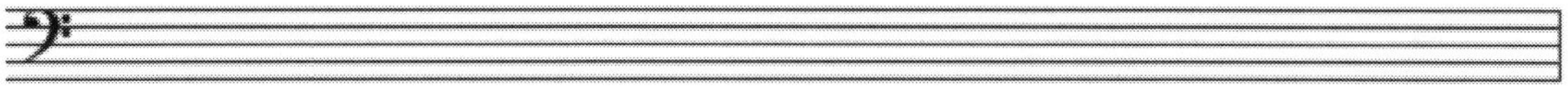

Middle C F D B C E G D E G

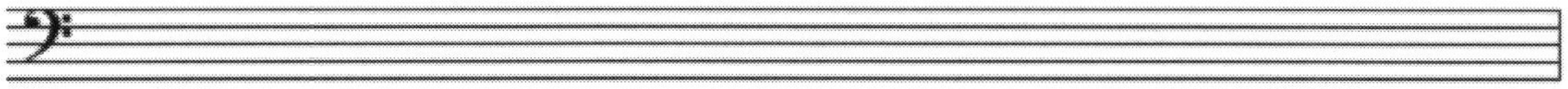

F C B D Middle C G F A G E

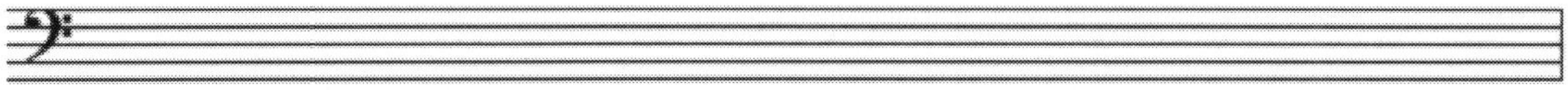

A D E F B G E C D F

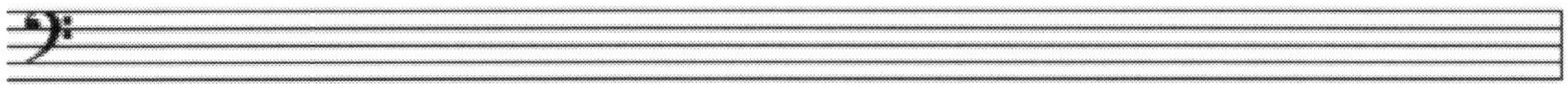

D C B E C B C D A G

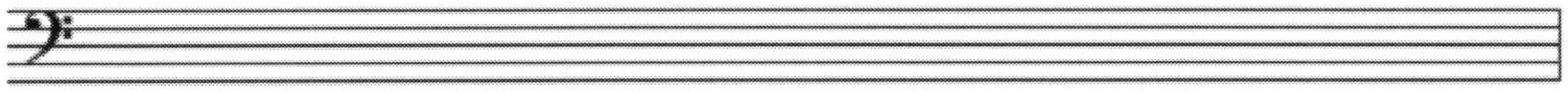

B D F E G A F A B Middle C

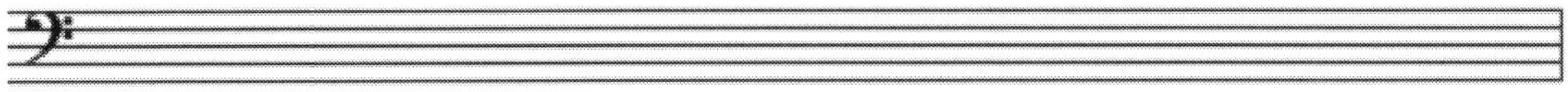

G C A D E G B F G E

Lesson 14

Treble and Bass Clef Notes on the Grand Staff

Note: The *Grand Staff* is the Treble and Bass Clef connected together

Treble and Bass Clef Exercise

Directions: Write the name of the note under each treble or bass clef note. **Try not to use the chart on the previous page.**

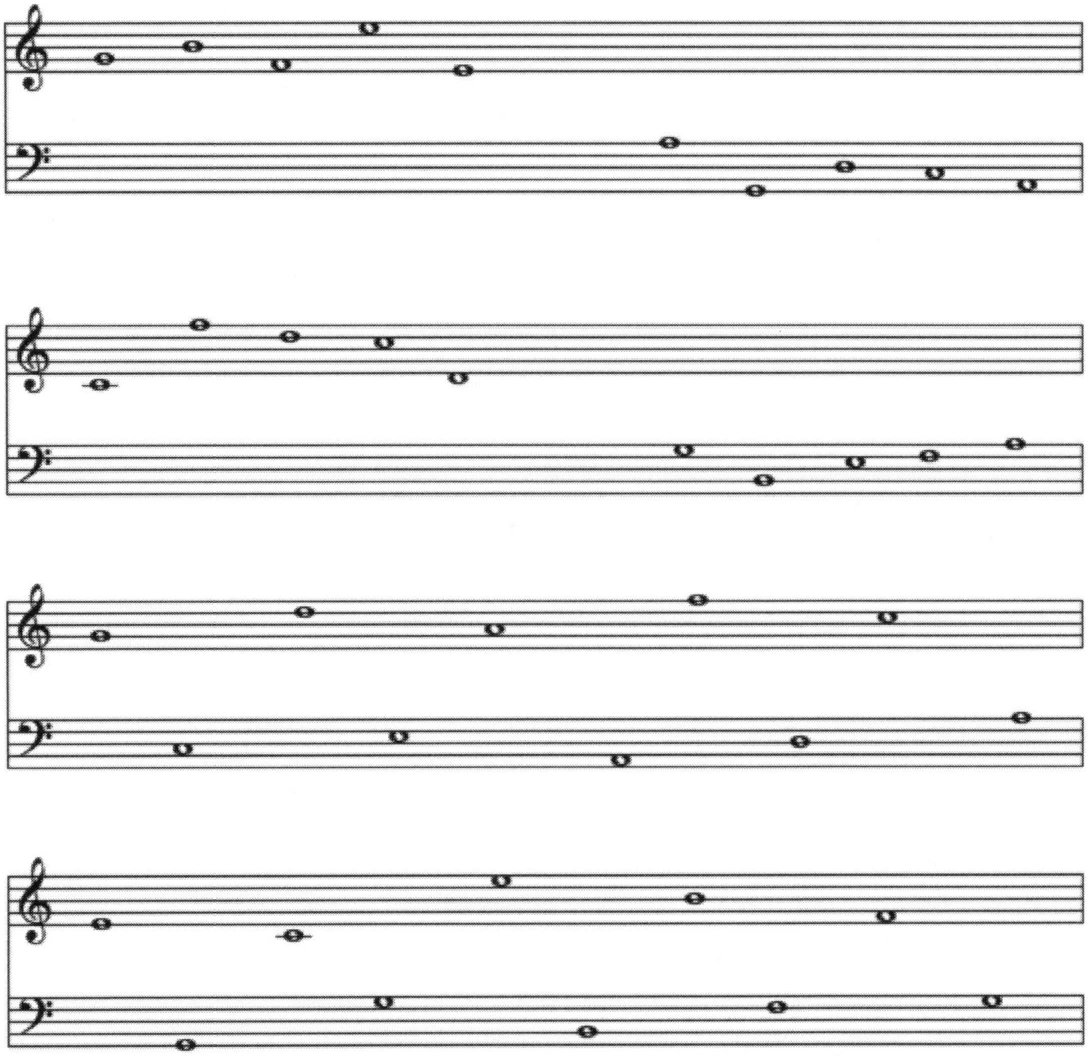

Treble and Bass Clef Exercise II

Directions: Recite the following notes in your head until you are able to read them without looking at the chart. Recite them backwards, forwards and out of order until you recognize each note. **Do not write the letter names of the notes.**

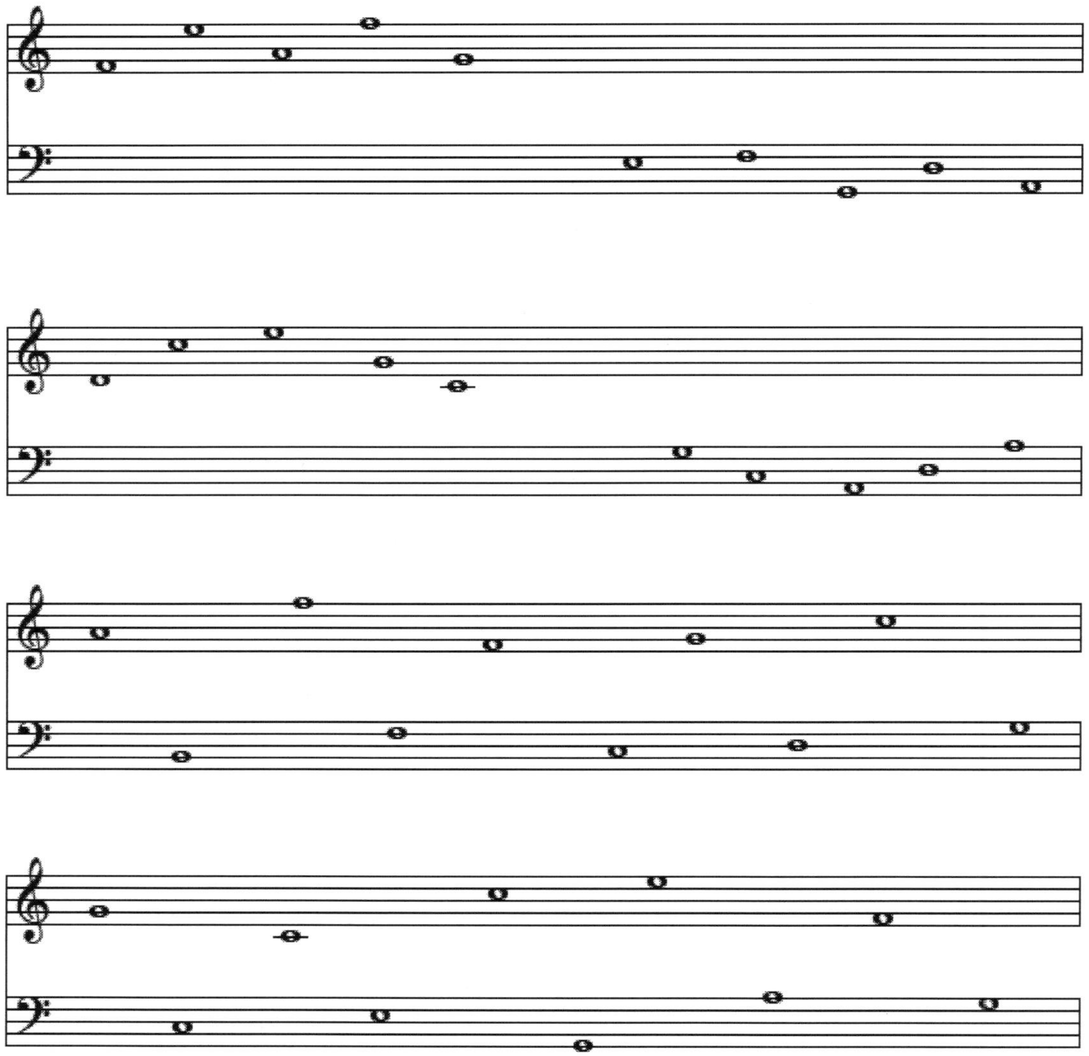

Lesson 15

Learning Time Signatures

In order to indicate rhythm in music, a **time signature** is used. A time signature consists of two numbers placed at the beginning of the staff. The top number indicates the number of beats within a measure, and the bottom number tells which kind of note equals one beat. Each measure **must** have the number of beats stated by the top number of the time signature.

(Track 24)

2 (2 beats in every measure)
4 (the quarter note equals 1 beat

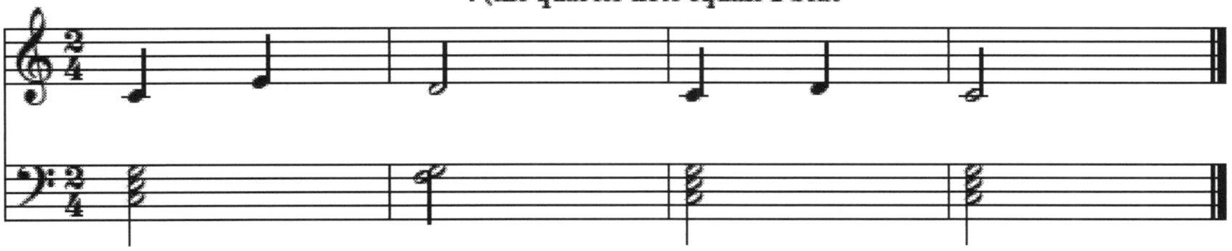

(Track 25)

3 (3 beats in every measure)
4 (the quarter note equals 1 beat

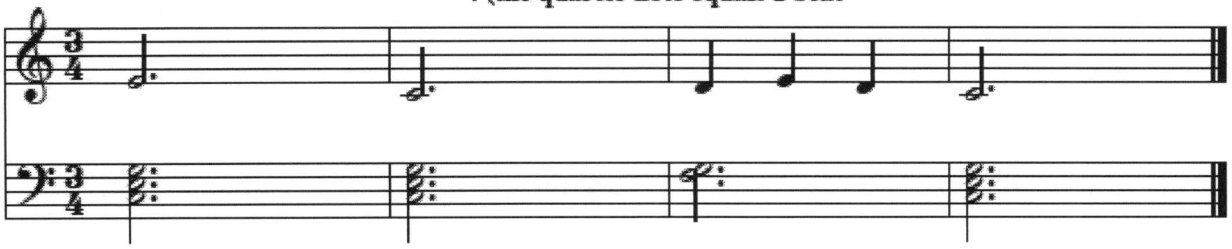

(Track 26)

4 (4 beats in every measure)
4 (the quarter note equals 1 beat

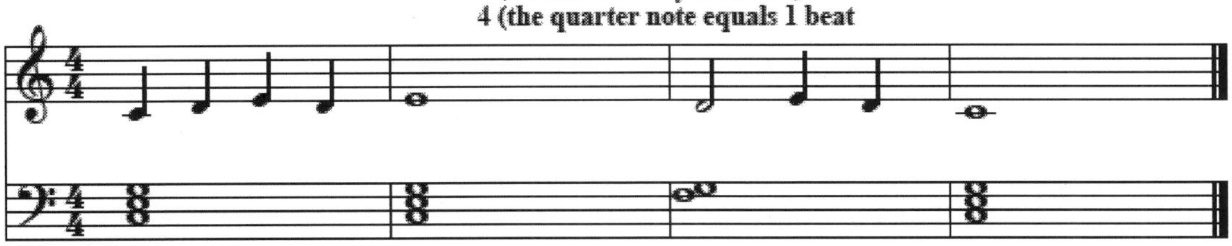

Time Signature Exercise

Directions: Place the correct time signature at the beginning of each exercise.

Exercise #1

Exercise #2

Exercise #3

Lesson 16

Learning Rests

A *rest* is a period of silence. Like notes, rests have values. Below are three different rests used in music. Notice there is no rest to indicate three beats. A three beat rest is notated by using a half rest plus a dot, a half rest plus a quarter rest or three quarter rests.

Quarter Rest = 1 beat

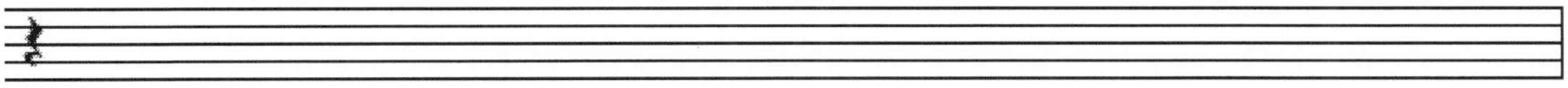

Half Rest = 2 beats

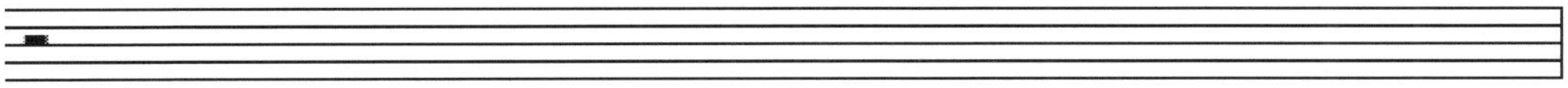

Whole Rest = 3 beats or entire measure when using a time signature other than $\frac{4}{4}$ time

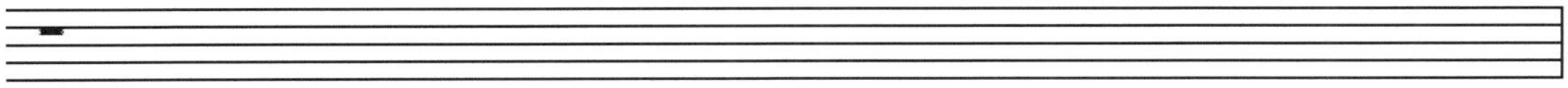

Rest Exercise

Directions: Figure out the time signature of each exercise and write it at the beginning of the staff after the treble clef. Then clap out the following rhythms. When you come to a rest, hold your hands apart for the value of the rest. When you come to a note that is longer than one beat, hold your hands together for the value of that note.

Exercise #1

Exercise #2

Exercise #3

Time Signature and Rest Exercises

Directions: Add barlines to create measures that equal the number of beats stated in the time signature.

Directions: Create four measures on each staff that equal the number of beats stated in the time signature by adding barlines, notes and rests.

Listening/Sightreading Exercise Resting (Track 27)

Directions: Play the following two songs on the piano or keyboard. (If you don't have either, guide your fingers on the piano chart.) Once you're comfortable playing the songs, make up words to the songs using one syllable per note to create you own song. **Don't ignore the rests and be sure to hold each note for its correct value.**

A Rest to Remember (Track 28)

Note Value Exercise

Directions: Write notes on the lines that add up to the sum given. Do not repeat a note more than two times. (Refer to Lesson 3 for help)

1. _____+_____ = 3 beats

2. _____+_____ = 2 beats

3. _____+_____ = 4 beats

4. _____+_____ + _____ + _____ = 7 beats

5. _____+_____+ _____ + _____+ _____ = 14 beats

6. _____+_____ + _____ + _____+_____+_____+_____+_____ = 20 beats

7. _____+_____+_____+_____ = 10 beats

8. _____+_____+_____+_____+_____ = 15 beats

9. _____+_____+_____+_____+_____+_____ = 12 beats

10. _____+_____+_____+_____+_____+_____+_____ = 16 beats

11. _____+_____+_____ = 9 beats

12. _____+_____+_____+_____ = 8 beats

13. _____+_____+_____ = 5 beats

14. _____+_____+_____+_____ = 11 beats

15. _____+_____+_____+_____+_____+_____ = 17 beats

Rest Value Exercise

Directions: Write rests on the lines that add up to the sum given. Do not repeat a rest more than two times.

Note: A whole rest equals four beats in this exercise. (Refer to Lesson 16 for help)

1. _____+_____ = 2 beats
2. _____+_____ = 3 beats
3. _____+_____+_____ = 8 beats
4. _____+_____ = 6 beats
5. _____+_____+_____ = 5 beats

Note and Rest Value Exercise

Directions: Write notes and rests on the lines that add up to the sum given. Do not repeat a note or rest more than two times.

Note: A whole rest equals four beats in this exercise.

1. _____+_____+_____+_____ = 5 beats
2. _____+_____+_____ = 9 beats
3. _____+_____+_____+_____+_____ = 18 beats
4. _____+_____+_____+_____+_____+_____ = 15 beats
5. _____+_____+_____+_____ = 16 beats
6. _____+_____+_____+_____+_____ = 8 beats
7. _____+_____+_____+_____+_____+_____ = 14 beats
8. _____+_____+_____ = 10 beats
9. _____+_____+_____+_____ = 7 beats
10. _____+_____+_____+_____+_____ = 11 beats

Music Word Find
After finding a word, draw the symbol that matches the word.

C	H	O	L	B	A	S	C	F	F	A	T	S	M	Q	F	D	Q
R	E	P	E	N	I	L	R	A	B	E	L	B	U	O	D	L	U
E	T	O	N	E	L	O	H	W	A	H	S	A	T	W	R	I	A
P	O	G	L	D	O	U	B	A	R	L	Z	S	Z	O	O	J	R
E	N	W	H	O	T	S	V	K	L	P	Q	S	H	V	H	U	P
C	F	H	A	L	K	R	J	J	I	R	U	C	L	T	C	T	S
T	L	B	A	S	L	E	F	Z	N	K	E	L	S	B	E	R	T
G	A	Q	S	E	R	U	S	A	E	M	O	E	X	P	Q	S	A
J	H	D	O	T	T	N	O	T	S	E	R	F	L	A	H	F	P
U	D	Q	U	A	R	T	E	R	R	E	S	T	J	D	G	W	H
Y	E	E	G	F	E	L	C	E	L	B	E	R	T	X	Z	S	K
A	T	C	B	V	D	E	T	O	N	R	E	T	R	A	U	Q	L
E	T	O	N	F	L	A	H	N	V	L	Y	K	H	F	D	B	U
B	O	F	R	B	T	W	Q	S	M	X	P	L	O	N	K	Q	M
V	D	C	W	Z	U	A	Z	N	G	I	S	T	A	E	P	E	R

Bar Line
Bass Clef
Chord
Dotted Half Note
Double Bar Line
Half Note
Measure
Quarter Note
Quarter Rest
Repeat Sign
Staff
Treble Clef
Whole Note
Whole Rest

Lesson 17

About the Drums

A drum is a musical instrument that dates as far back as the Medieval Period with the introduction of the kettledrum, which is made up of a metal bowl covered with a membrane stretched tightly over the opening. Before the kettledrum, rattles and clappers (sticks struck together) were used to create rhythms during the Ancient period.

The drum is categorized as a percussive instrument because it is struck with another object, such as sticks or the hands, to produce its sound. There are many types of drums but the ones that are most familiar are the ones found in a seven-piece drum set.

Like that of the kettledrum, the drums found in a drum set consist of a frame that has one or more membranes or skins stretched over it. The sound of the drum is created when the membranes are struck and the frame resonates. The sound of the tom toms, floor tom and bass drums (see diagram) are produced this way. The sound of the snare drum is a little different because in addition to the stretched skin, it also consists of gut strings wound with wire on the bottom of it. This causes the drum to rattle when it is struck.

The other two pieces of the drum set are the hi hat and the crash cymbal. The hi hat consists of two metal plates that strike each other when its pedal is pressed. The crash cymbal is one metal plate that is struck by a drum stick. When all of the pieces of the drum set are played together in a logical manner, various rhythms with a uniform blend of alternating beats are created.

Pieces of a drum set:

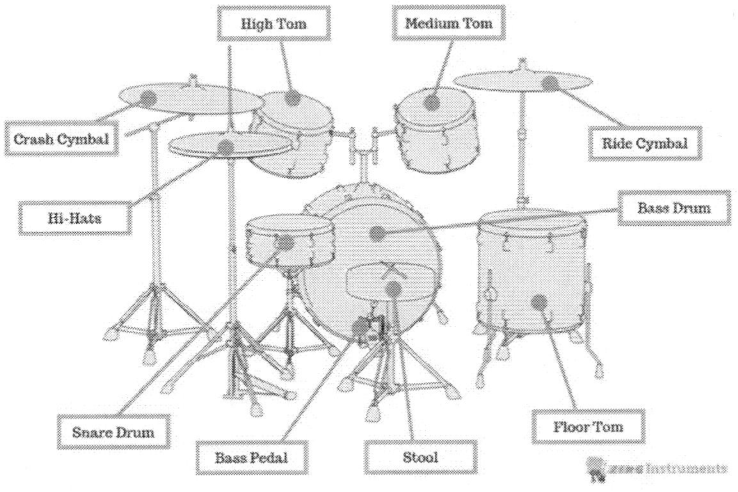

Drum Exercise: Directions: Name the pieces of the drum set.

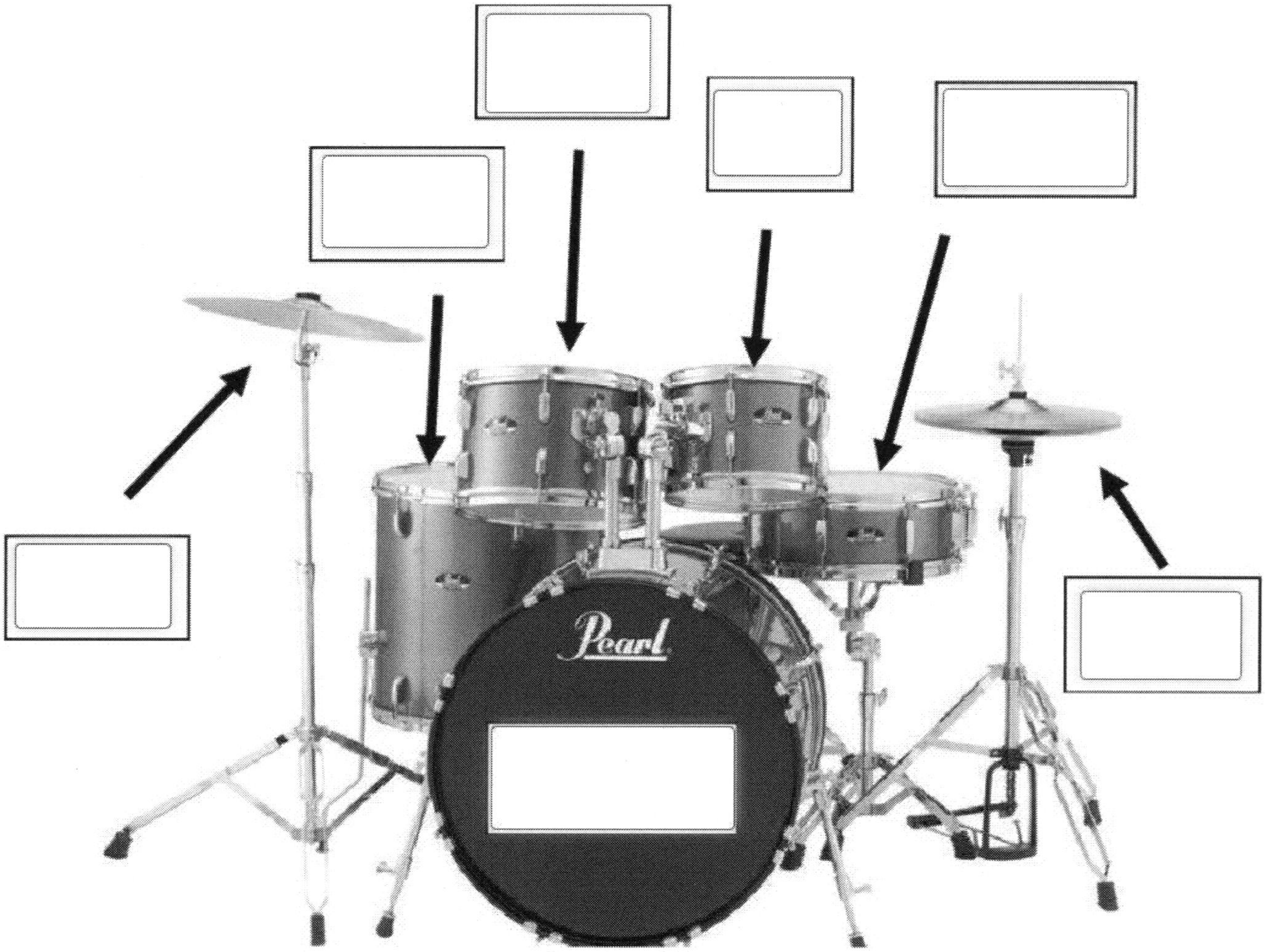

Learning Duple and Triple Meter (Track 29)

Duple Meter: any rhythm that is divisible by two

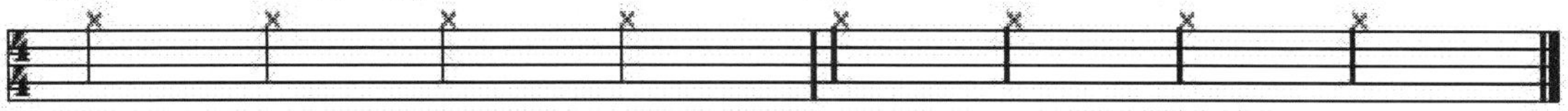

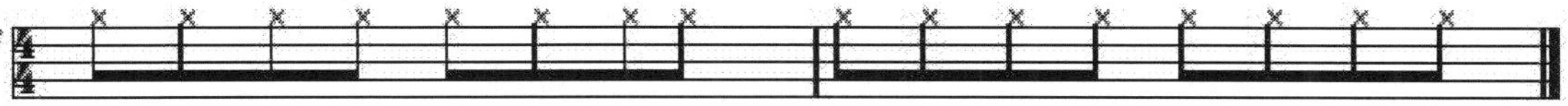

(Track 30)

Triple Meter: any rhythm that is divisible by three.

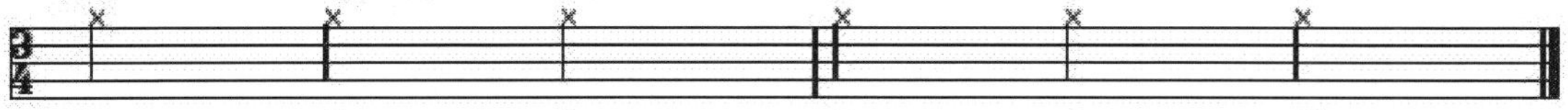

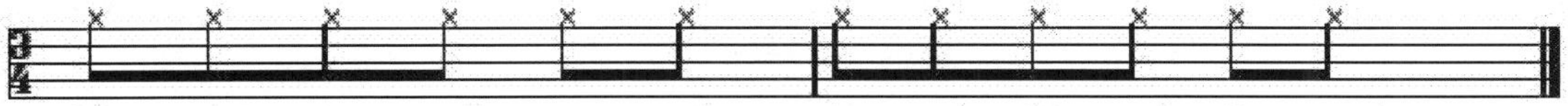

***Note:** The beamed (connected) notes are eighth notes which are twice as fast as eighth notes. If quarter notes equal one beat, then eighth notes equal one-half of a beat.

Rhythmic Exercise: Duple meter (Track 31)

Directions: Clap the following rhythms in duple meter until you clap the rhythms comfortably. Count out loud while clapping.

Rhythm #1

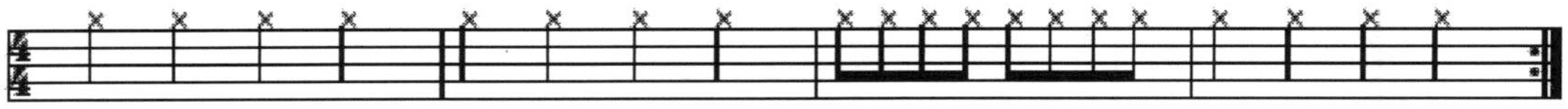

1 2 3 4 1 2 3 4 1 & 2 & 3 & 4 & 1 2 3 4

Rhythm #2

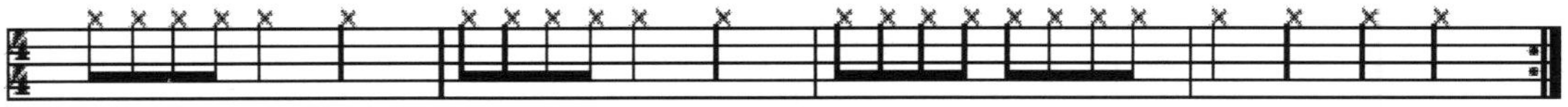

1 & 2 & 3 4 1 & 2 & 3 4 1 & 2 & 3 & 4 & 1 2 3 4

Rhythm #3

1 2 1 & 2 & 1 & 2 & 1 2

Rhythm #4

1 2 & 1 2 & 1 2 & 1 2

Directions: Clap out the following rhythms in triple meter until you clap the rhythms comfortably. Count out loud while clapping.

Rhythm #1

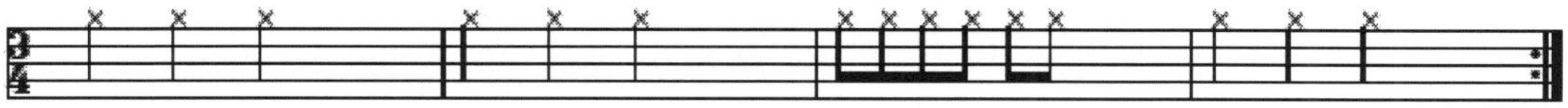

1 2 3 1 2 3 1 & 2 & 3 & 1 2 3

Rhythm #2

1 & 2 & 3 1 & 2 & 3 1 & 2 & 3 & 1 2 3

Rhythm #3

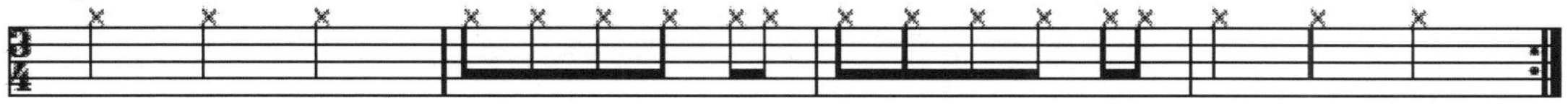

1 2 3 1 & 2 & 3& 1 & 2 & 3 & 1 2 3

Rhythm #4

1 2 & 3 1 2 & 3 1 2 & 3 1 2 3

Rhythmic Exercise: Triple Meter (Track 32)

Rhythmic Exercise: Duple and Triple Meter

Directions: Learn the two sets of rhythms in duple meter and tap them on a desk or table. Once you are comfortable tapping both rhythms, tap rhythm #1 while a fellow music student taps rhythm #2, then switch rhythms. If you're in a music class, the teacher should separate the class into two groups and instruct one group to tap rhythm #1 and the other group to tap rhythm #2 separately. Once the groups are comfortable tapping their rhythms, then the teacher should instruct them to tap their rhythms simultaneously. Then the groups should be instructed to switch rhythms. Repeat the same exercise with the two sets of rhythms in triple meter.

Duple Meter
(Track 33)

Rhythm #1

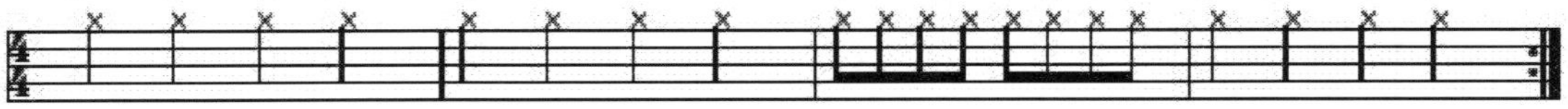

1 2 3 4 1 2 3 4 1 & 2 & 3 & 4 & 1 2 3 4

Rhythm #2

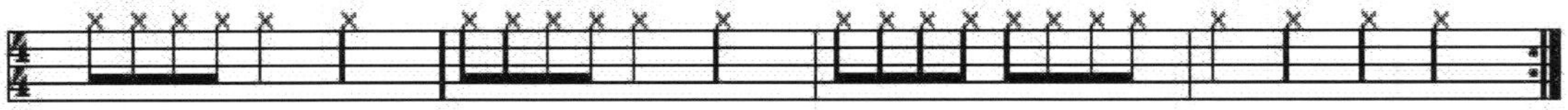

1 & 2 & 3 4 1 & 2 & 3 4 1 & 2 & 3 & 4 & 1 2 3 4

Triple Meter
(Track 34)

Rhythm #1

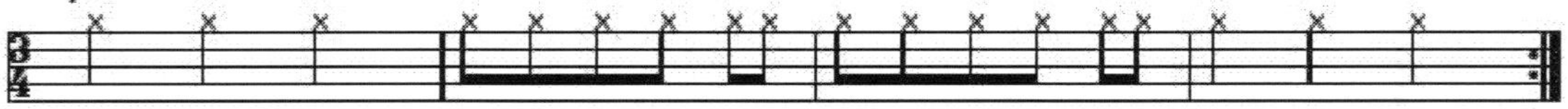

1 2 3 1 & 2 & 3 & 1 & 2 & 3 & 1 2 3

Rhythm #2

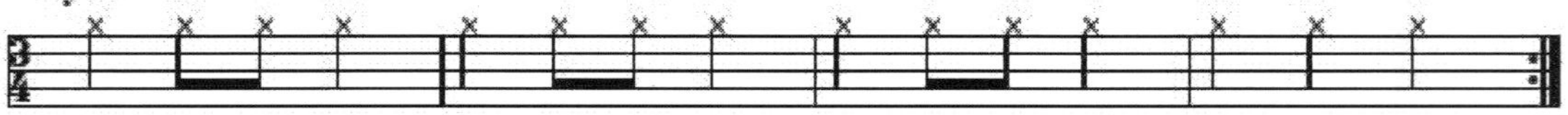

1 2 & 3 1 2 & 3 1 2 & 3 1 2 3

Creative Rhythm Exercise

Directions: Write out two rhythms in duple meter and two rhythms in triple meter and then have a fellow music student or your music class clap out your rhythms.

Duple Meter

Rhythm #1

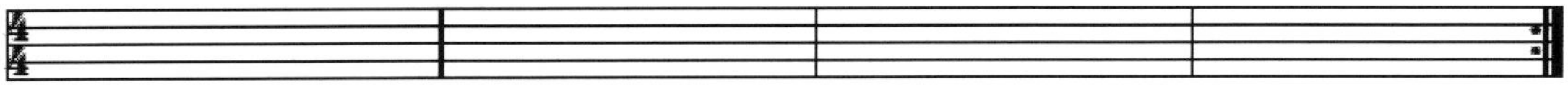

Rhythm #2

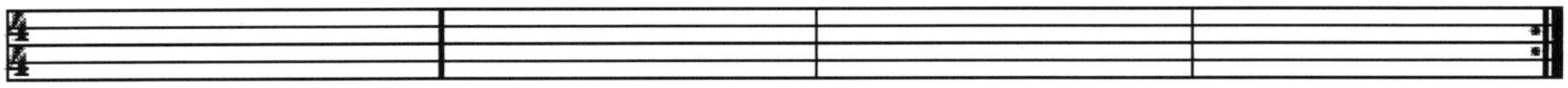

Triple Meter

Rhythm #1

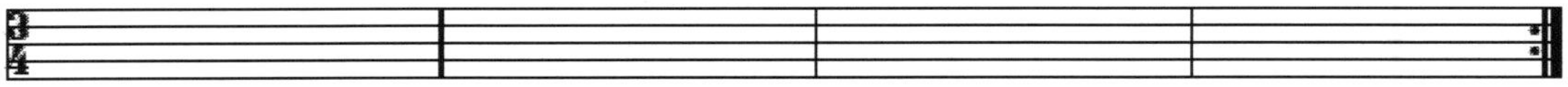

Rhythm #2

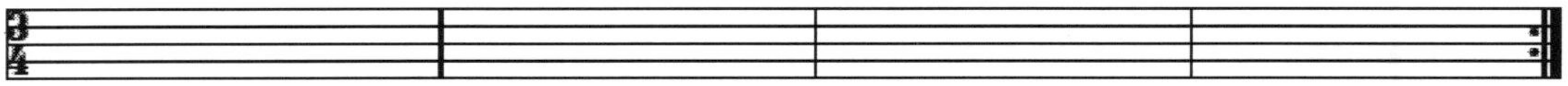

Drum Exercise

Directions: Try to play the following rhythms in duple and triple meter on a drum set using the hi hat, snare and bass drum. Pay close attention to the notation.

Duple Meter
(Track 35)

Rhythm #1

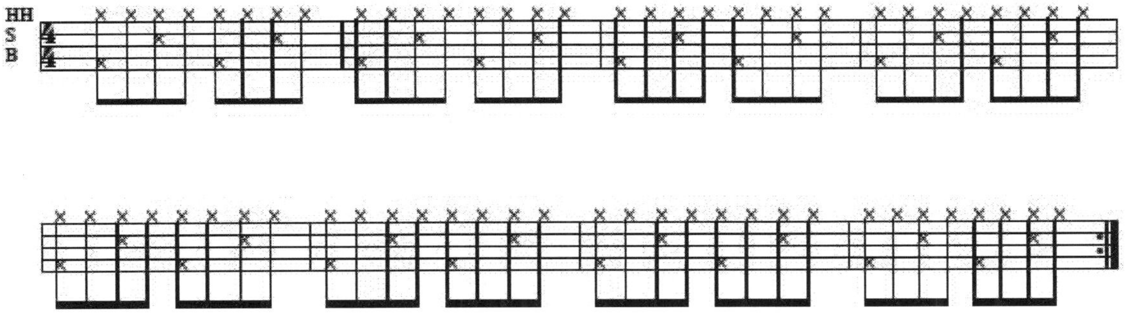

Triple Meter
(Track 36)

Rhythm #2

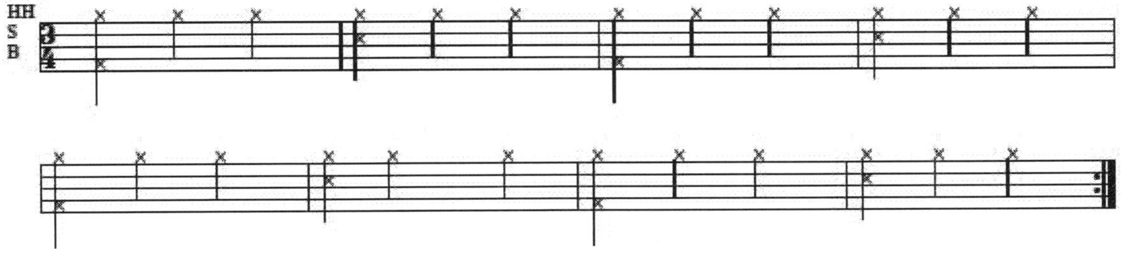

HH = Hi Hat (Right Hand)
S = Snare Drum (Left Hand)
B = Bass Drum (Right Foot)

Music Word Find

T	O	M	S	B	A	S	R	U	C	H	E	D	K	M
I	Q	R	H	Y	T	Z	E	L	P	I	R	T	E	D
M	T	C	R	A	S	N	A	G	L	H	F	O	U	P
E	O	W	X	M	U	R	D	S	S	A	B	P	J	Q
E	M	I	T	R	U	O	F	O	W	T	L	H	I	H
R	T	M	O	T	R	O	O	L	F	E	D	U	P	L
H	O	V	E	M	I	T	R	U	O	F	R	U	O	F
Y	M	V	R	E	T	E	M	E	T	O	L	A	K	M
T	S	E	T	O	N	H	T	H	G	I	E	V	N	R
T	I	M	E	S	I	G	N	A	T	U	R	E	O	S
F	O	U	S	T	I	M	L	B	D	Y	C	R	A	S
E	M	I	T	R	U	O	F	E	E	R	H	T	L	J
R	R	Y	L	A	B	M	Y	C	H	S	A	R	C	Z

Bass drum
Crash Cymbal
Duple
Eighth Notes
Floor tom
Four-four time
Hi hat
Meter

Rhythm
Snare
Three-four time
Time Signature
Tom toms
Triple
Two-four time

Lesson 18
Music Set to Poetry

Directions: Write a poem that will be set to music. The poem should be a **positive** rhyme or rap, a poetic letter or a non-rhyming poem.

Lesson 19:
Poetry Set to Music

Directions: Write a melody and add accompaniment (chords). Set this music to your original poem created in the previous chapter. Refer to lessons 1- 17 for help. When the song is completed, record it. **Make sure the music has the same feeling as your poem**.

Creative Song #1
Directions: Write a melody and accompaniment to create your own song.

1. ______________________________

Directions: Create a melody that fit with the chords in the bass clef.

Directions: Write a melody and accompaniment to create your own song.

2.____________________________________

Directions: Write a melody and accompaniment to create your own song.

3.________________________________

Directions: Write a melody and accompaniment to create your own song.

4.____________________________________

Lesson 20

Welcome to the World of Music History

In order to appreciate music completely, its history should be understood. This section will explain how music reflected ideas and events of important periods of European and forms of American history. It begins with a list of suggested listening material, which helps to clarify the information discussed. Also, it includes instruments and key composers that dominated the music of each era. The first half of this section pertains to European music history and the second half discusses the progressions of Jazz and other forms of African American music.

It is important to note that music progressed and changed gradually; therefore, the dates for each historical time frame are approximations. In other words, while a new style of music might have been influenced by a previous style, the previous style didn't just vanish when the new style was created. The previous style might have remained for another ten, twenty or more years while the new style was developing.

History plays a major role in the development of music. Its effect on music is key to understanding the progression of music. By examining the events, instruments and composers of each era, it becomes quite clear as to how music was affected by social issues, and progressed to the music that we study, understand and listen to today.

The Progression of European Music with Suggested Listening Material

Ancient Period:

Euripides *Orestes, Stasimon chorus*
Epitaph of Seikilos

Medieval Period:

Leonin: *Alleluia Pascha nostrum* (Organum duplum (two voices))
Perotin: *Sederunt* (Organum quadruplum: (four voices))
Guillaume de Machaut: *Quant Theseus-Ne quier veoir* (Double Ballade)
Salve, sancta parens (Carol)

Renaissance Period:

Dufay: *Se la face ay pale* (Ballade)
Heinrich Isaac: *Innsbruck, ich muss dich lassen* (Lied)
Josquin des Prez: *Tu Solus, qui facis mirabilia* (Motet)
Giovanni da Palestrina: *Pope Marcellus Mass: Credo* (Mass)

Baroque Period:

Antonio Vivaldi: *Four Seasons*
J.S. Bach: *Orchestral Suite No. 3: Air on a G String; Prelude in C Minor; Fugue in G Minor*
George Frederick Handel: *Messiah: Hallelujah Chorus; The Water Music: Suite No. 2*

Classical Period:

Wolfgang Amadeus Mozart: *Serenade No. 10 in B Flat Major;* Clarinet Concerto in A Major
Ludwig van Beethoven: *Symphony No. 6 (Pastoral Symphony); Romance No. 1 for Violin and*
Orchestra; Moonlight Sonata
Joseph Haydn: *Serenade*

Romantic Period

Franz Schubert: *Serenade; Ballet Music in G Major from Rosamunde*
Robert Schumann: *Dichterliebe: In wunderschonen Monat Mai; Kinderszenen: Traumerei*
Johannes Brahms: *Symphony No. 3 Poco allegretto; String Quintet in G: Adagio*
Franz Liszt: *Liebstraum No. 3: Poco allegretto; Consolation No. 3 in D Flat Major*
Frederic Chopin: *Symphony No. 9 in E Minor, Op. 95 From the New World*
Johann Strauss: *The Blue Danube Waltz; Roses from the South and The Radetsky March*

Twentieth Century
Igor Stravinsky: *Rite of Spring: Petrushka*

Impressionism

Claude Debussy: Prelude a l'apres-midi d'un faune; La Mer
Maurice Ravel: Jeux d'eau; Daphnis and /chloe: Nocturne

Expressionism

Arnold Schoenberg: Five Pieces for Orchestra, Op. 16; Six Little Piano Pieces

Music during the Ancient Period (app. 753 B.C. – 336 A.D.)

Music of the Ancient Greeks

Historical Background: Music during this period reflected a time when the Greeks believed that "music had a direct effect on the will and thus on the character and conduct of human beings." This doctrine is known as the *doctrine of ethos.*[1] Aristotle, a Greek philosopher, had a similar view. He believed in what was called the *doctrine of imitation*, which states "music directly imitates the passions or states of the soul – gentleness, anger, courage, temperance, and their opposites and other qualities; hence, when one listens to music that imitates a certain passion, one becomes imbued with the same passion…"[2]

About the Music: Contrary to what people might think, the Greek musical system was quite sophisticated for that time. It consisted of notes, scale systems known as modes and compositions based on melody.

The modes represented different emotions because of their sound. For instance, it was believed that the *Dorian* mode produces a relaxed temper while the *Phrygian* mode produces enthusiasm.

Melody was very important at this time because it made up an entire composition. There was no harmony (accompaniment or chords). This type of music is called *monophony.*

Instruments:

- <u>Lyre</u> – a stringed instrument, consisting of five to seven strings, at first, then, possessing eleven strings. It is used for solo playing and accompanying singing or reciting epic poems.
- <u>Kithara</u> – the larger counterpart of the lyre.
- <u>Aulos</u> – a single or double-reed instrument (not a flute) that makes a shrill tone and is used to play with certain types of poetry used for worshipping Greek Gods.

Music of the Ancient Romans

Historical Background: It is believed that Rome took its art music from Greece, especially after that country became a Roman province in 146 B.C.

Instruments:

- Tibia – the Roman version of the aulos. It played an important role in religious rites, military music and the theater.
- Tuba – a long, straight trumpet derived from the Etruscans. It is used in religious, state and military ceremonies.
- Cornu – a G-shaped circular horn
- Buccina – a smaller version of the cornu

---

[1] Donald J. Grout and Claude V. Palisca, *A History of Western Music* (New York, 1988), p.7.

[2] Ibid. pp. 7-8.

Quiz 1

1. What are the approximate dates of this period?

2. What was the historical background of the music of this period? Explain the thoughts of the people or any major events taking place that had a direct effect on the music during this time period.

3. What type of music flourished during this time period?

4. In what country (ies) did the music of this time flourish?

5. Who composed the music of this period?

6. Name some of the instruments used during this time period.

7. Listen to the music of this time period and describe what you hear. List some characteristics.

8. Listen to the music of this time period and explain what comes to your mind when you hear it.

9. Listen to the music of this time period and describe how it makes you feel when you hear it and explain what causes you to feel that way. For example, do you think it's the instruments, the tempo, the words or something else that you hear that causes you to feel the way that you do?

10. If you are in a classroom setting, listen to a particular song of this time period and create a scene **without words** that fits the music. Write it down then act out the scene. Have the class attempt to figure out the meaning of your scene. If you are studying music individually, listen to a particular song of this time period, create a story that fits the music. Record it in a journal.

Music during the Medieval Period (app. 336-1500)

Historical Background: Boethius, considered the most influential authority on music in the Medieval Period (Middle Ages), believed that "the true musician is not the singer, or one who only makes up songs by instinct without knowing the meaning of what he does, but the philosopher, the critic, he who exhibits the faculty of forming judgments according to speculation or reason appropriate to music…"[3]

About the Music: By the middle of the Medieval Period, music began to take shape. From 600A.D. –1000 A.D. music was organized into notation. By 1000 A.D., *monophony* had given way to *polyphony*, which is music consisting of a melody being doubled by another pitch that was either a third, fourth or fifth above the melody. This type of music is called *organum*.

In addition to the development of polyphony, the development of *secular* music (music not about God or the church) took place and became more popular than *sacred* music (music about God). However, both types of music became more refined.

Instruments:

Low	High	Percussive
Harps	Cornets	Kettle drums
Lutes	Slide trumpets	Small bells
Psalteries (string instruments)	Sackbuts (early form of trombone)	Cymbals
Portative Organ		
Flutes		
Recorders		

Composers:

- Leonin composed organum using two voices
- Perotin composed organum using three and four voices
- Guillaume de Machaut

[3] Ibid. p.39.

Sacred Music in Rome

Sacred music in Rome was used in two types of religious services. One service was called the *Office* and the other service was the *Mass*. During the Office, psalms and scriptures from the Bible were performed as chants, daily.
Chanting is a cross between singing and speaking. During the Mass, the Last Supper (the last meal Jesus Christ ate before He was crucified on the cross) was re-enacted and accompanied by a choir and congregation singing in two voices.

Secular Music in France

In France, *secular* music began to flourish. For instance, the *motet* was a type of composition that consisted of three or more voices (like that of Perotin's organum), with each voice having a different text. This probably sounds confusing but the music was unified because the subject of the text was the same. In other words, the text of all three voices was about one subject: love, nature or a subject not related to God or the church. Another type of music that developed in France was the *hocket*. The hocket contained flowing melodies that were interrupted by rests, creating a "hiccup" sound.
The thirteenth century (1200's) was an important time for, both, sacred and secular music because it was during this time that music notation was refined. It was possible for future generations to examine and understand more clearly the preserved music of the past.

The Flourishing of French Music of the Fourteenth Century (1300's)

During the fourteenth century in France, the authority of the church was questioned and the French monarchy was strengthened. Therefore, the music of this time reflected an even more secular and less sacred world. Also, the music was more unified harmonically because melodies were doubled by pitches a third or a sixth above (similar to the method used in *organum*). *Guillaume de Machaut's ballade* is a good example of this. His ballade consists of a vocal line accompanied by two instrumental lines.

The Flourishing of Italian Music of the Fourteenth Century

During the fourteenth century, Italy's city-states constantly rivaled for ruling power. While this went on, polyphony grew in elite circles as a secular entertainment. Out of this, the *caccia* developed. The caccia consists of two voices that sang the same melody (mixed texts) but at different times. One voice followed the other or "chased" the other. Both, the music of the French and the Italian possessed rhythmically free characteristics.

The Flourishing of English Music of the Fifteenth Century (1400's)

English music in the fifteenth century had a structure of its own. It tended to be, what we would call today, "more chordal." This means that it didn't have the independent lines like that of the *caccia* however, it was closer in structure to the ballade only it had more harmonic unity and the rhythm was less free. The *carol* is an example of this style.

Quiz 2

1) What are the approximate dates of this period?

2) What was the historical background of the music of this period? Explain the thoughts of the people or any major events taking place that had a direct effect on the music during this time period.

3) What type of music flourished during this time period?

4) In what country (ies) did the music of this time flourish?

5) Who composed the music of this period?

6) Name some of the instruments used during this time period.

7) Listen to the music of this time period and describe what you hear. List some characteristics.

8) Listen to the music of this time period and explain what comes to your mind when you hear it.

9) Listen to the music of this time period and describe how it makes you feel when you hear it and explain what causes you to feel that way. For example, do you think it's the instruments, the tempo, the words or something else that you hear that causes you to feel the way that you do?

10) If you are in a classroom setting, listen to a particular song of this time period and create a scene **without words** that fits the music. Write it down then act out the scene. Have the class attempt to figure out the meaning of your scene. If you are studying music individually, listen to a particular song of this time period, create a story that fits the music. Record it in a journal.

Music during the Renaissance Period (app. 1450-1600)

Historical Background: Music at this time reflected a movement called *humanism*, which is the revival of ancient learning, particularly of grammar, rhetoric, poetry, history and moral philosophy. It was a "rebirth" of the classics. Also during this time, music printing was born.

About the Music: Music was still polyphonic and contained independent voices, but the composers during this time were more interested in their music sharing the same sentiment or feeling as the meaning of the text it accompanied. This is called *musica reservata.*[4]

Instruments:

- Wind instruments
- Church organ
- Harpsichord – a keyboard instrument used for solo or ensemble music
- Clavichord – a keyboard instrument, softer that the harpsichord, used for solo music
- Lute
- Harp
- Kettledrum

Composers:

- Dufay
- Josquin des Prez
- Henrich Isaac
- Giovanni da Palestrina

Note: Polyphony in Germany and Spain developed later than in France, Italy and England. In 1530, the German *lied* was born and in the 1490's, the Spanish *villancico* was created.

Sacred Music in Germany during the Renaissance

The *Lutheran Chorale* developed after Martin Luther, a philosopher then monk and choirmaster, became leader of a movement that protested against the Church of Rome. This was called the *Protestant Movement.*

---

[4] Ibid. p.231.

The Lutheran Chorale was a congregational hymn consisting of four voices: the top voice was the melody and the lower three voices were the accompaniment. This style of music is very similar to the hymn, as we know it today.
Some of the melodies were borrowed from secular songs such as love songs and were sung by choirs accompanied by an organ or even an orchestra.

Sacred Music Outside of Germany during the Renaissance

Another type of music that was used for worship was the *psalter*. The psalter developed out of the Calvinist Church led by John Calvin, a leader of a "reformed" Protestant sect. He opposed, completely, the singing of texts not found in the Bible. The psalter is a psalm set to music. Like the chorale, this music consisted of four or more parts with the melody occupying the top or third voice.

Composer: Because there appeared to be a need to rid the church of anything deemed inappropriate within the church, a council was put together at Trent in Northern Italy in 1545. This council set the tone for the church. With respect to music, the main complaint was that the music of the church had a secular spirit due to polyphony. Therefore, a composer by the name of *Giovanni da Palestrina* composed a six-voice Mass that was very conservative in style (unlike the Lutheran Chorale). With this Mass, Palestrina proved that it is possible for polyphony to have a reverent and sacred spirit. It has been said that Palestrina was named "the savior of polyphonic church music."

Quiz 3

1) What are the approximate dates of this period?

2) What was the historical background of the music of this period? Explain the thoughts of the people or any major events taking place that had a direct effect on the music during this time period.

3) What type of music flourished during this time period?

4) In what country(ies) did the music of this time flourish?

5) Who composed the music of this period?

6) Name some of the instruments used during this time period.

7) Listen to the music of this time period and describe what you hear. List
   a. some characteristics.

8) Listen to the music of this time period and explain what comes to your mind when you hear it.

9) Listen to the music of this time period and describe how it makes you feel when you hear it and explain what causes you to feel that way. For example, do you think it's the instruments, the tempo, the words or something else that you hear that causes you to feel the way that you do?

10) If you are in a classroom setting, listen to a particular song of this time period and create a scene **without words** that fits the music. Write it down then act out the scene. Have the class attempt to figure out the meaning of your scene. If you are studying music individually, listen to a particular song of this time period, create a story that fits the music. Record it in a journal.

Music during the Baroque Period (app. 1600-1750)

Historical Background: Music during this period reflected a "flamboyant, decorative and expressionistic style."[5] It is believed that it took on the same qualities as that of architecture and art of the time period. However, it reflected tension that was brewing during this time as well. The beginning of the Baroque period was marked with the burning of the Italian philosopher, Giordono Bruno who was condemned for having beliefs that were contrary to the church in 1600. It was evident that the state of unrest between the sacred and secular worlds was increasing.

About the Music: The music consisted of a strong continuous bass (bottom voice), a decorative treble (top voice) and a more subtle harmony (middle voice or voices). Dissonant notes (notes that create tension) are introduced to the music and there are key changes within the compositions.

It was during this time that sacred and secular opera (drama set to continuous music) developed along with the growth of purely instrumental music. In fact, instrumental music became just as important as vocal music. Because of this importance, purely instrumental compositions such as the *toccata*, *fugue* and *prelude* were born.

Instruments:

- Organ
- Two-manual harpsichord
- Violin (and the string family)
- Brass instruments
- Harpsichord
- Flute
- Oboe
- Bassoon
- Trumpet
- Timpani

Composers:

- Antonio Vivaldi
- Johann Sebastian Bach
- George Frideric Handel

---

[5] Ibid. p. 346.

Quiz 4

1) What are the approximate dates of this period?

2) What was the historical background of the music of this period? Explain the thoughts of the people or any major events taking place that had a direct effect
    a. on the music during this time period.

3) What type of music flourished during this time period?

4) In what country(ies) did the music of this time flourish?

5) Who composed the music of this period?

6) Name some of the instruments used during this time period.

7) Listen to the music of this time period and describe what you hear. List some characteristics.

8) Listen to the music of this time period and explain what comes to your mind when you hear it.

9) Listen to the music of this time period and describe how it makes you feel when you hear it and explain what causes you to feel that way. For example, do you think it's the instruments, the tempo, the words or something else that you hear that causes you to feel the way that you do?

10) If you are in a classroom setting, listen to a particular song of this time period and create a scene **without words** that fits the music. Write it down then act out the scene. Have the class attempt to figure out the meaning of your scene. If you are studying music individually, listen to a particular song of this time period, create a story that fits the music. Record it in a journal.

Music during the Classical Period (app. 1750-1827)

Historical Background: Music during this period reflected a movement called the *Enlightenment*. This movement favored common sense and applied science over supernatural religion.

About the Music: The music of this time is set up into movements that contrast in moods, texture and keys to create drama and expression. The *sonata* is a prime example of this mood music. It begins with two contrasting musical ideas; goes into what is called a "development," which unfolds the two ideas, and then returns to the opening material. The melodies and rhythms are put together in a very logical and organized manner.

Instrumental Music:

- Sonata – written for two, three, four or five instruments. It contains three or four movements of contrasting mood and *tempo* (rate of speed).
- Symphony – a sonata written for an orchestra
- Concerto – an orchestral sonata with a soloist

Composers:

- Franz Haydn
- Wolfgang Amadeus Mozart
- Ludwig Van Beethoven

Quiz 5

1) What are the approximate dates of this period?

2) What was the historical background of the music of this period? Explain the
   a. thoughts of the people or any major events taking place that had a direct effect
   b. on the music during this time period.

3) What type of music flourished during this time period?

4) In what country(ies) did the music of this time flourish?

5) Who composed the music of this period?

6) Name some of the instruments used during this time period.

7) Listen to the music of this time period and describe what you hear. List
   a. some characteristics.

8) Listen to the music of this time period and explain what comes to your mind when you hear it.

9) Listen to the music of this time period and describe how it makes you feel when you hear it and explain what causes you to feel that way. For example, do you think it's the instruments, the tempo, the words or something else that you hear that causes you to feel the way that you do?

10) If you are in a classroom setting, listen to a particular song of this time period and create a scene **without words** that fits the music. Write it down then act out the scene. Have the class attempt to figure out the meaning of your scene. If you are studying music individually, listen to a particular song of this time period, create a story that fits the music. Record it in a journal.

Music during the Romantic Period (1827-1900)

Historical Background: Music during this time reflected "freedom, movement, passion and an endless pursuit of the unattainable."[6] The music was liberating because it was seen as a revolt against the limitations of the Classical period.

The music of the Romantic age was influenced by the support of political advancement (known as nationalism), which led to national pride. Different countries developed their own specific styles of music. Countries had their own characteristic styles before, but now composers made a conscious effort in developing their music to represent their nation. This was evident in Germany and France.

About the Music: Subjects that attract the Romantic composer are heroism, love and death. *Program Music* (instrumental music associated with poetic descriptive or narrative subject matter)[7] was developed during this time. The music is emotionally powerful to those who are touched by music that paints passionate pictures. The *oratorio* (a biblical drama) developed at this time.

Composers of German Romanticism:

- Franz Schubert
- Robert Schumann
- Johannes Brahms

Composer of French Romanticism:

- Franz Liszt
- Frederic Chopin

---

[6] Ibid. p.658.
[7] Ibid. p.660.

Quiz 6

1) What are the approximate dates of this period?

2) What was the historical background of the music of this period? Explain the
    a. thoughts of the people or any major events taking place that had a direct effect
    b. on the music during this time period.

3) What type of music flourished during this time period?

4) In what country(ies) did the music of this time flourish?

5) Who composed the music of this period?

6) Name some of the instruments used during this time period.

7) Listen to the music of this time period and describe what you hear. List
    a. some characteristics.

8) Listen to the music of this time period and explain what comes to your mind when you hear it.

9) Listen to the music of this time period and describe how it makes you feel when you hear it and explain what causes you to feel that way. For example, do you think it's the instruments, the tempo, the words or something else that you hear that causes you to feel the way that you do?

10) If you are in a classroom setting, listen to a particular song of this time period and create a scene without words that fits the music. Write it down then act out the scene. Have the class attempt to figure out the meaning of your scene. If you are studying music individually, listen to a particular song of this time period, create a story that fits the music. Record it in a journal.

Music during the Early to Mid-Twentieth Century (1900-1951)

Historical Background: The beginning of the twentieth century marked an emotional time, an uncertain time, and a time of social unrest. This tension was caused by World Wars I and II and it had a direct effect on the music of this time.

About the Music: During the late 1800's, a style of art and poetry that brought about moods or feelings through harmony and color was called *Impressionism.*[8] Impressionism is similar to *program music* (discussed in the section on Romantic music) in that it paints a picture but the picture is used to strike an emotion and not tell a story. Even though this term referred to art and poetry, there was a composer who captured this same image in his music. This composer was *Claude Debussy.* One of his famous works was the orchestral piece, *Prelude a l'apres-midi d'un faun.* This piece is based on a poem by Malarme and paints a picture of a mythical, half man, and half animal creature. In this piece, Debussy was not trying to portray his emotions. Instead he was trying to cause a reaction from his listeners.

Another composer, by the name of *Maurice Ravel,* composed music in the impressionistic style also but unfortunately, some critics labeled him an imitator of Debussy. Even though there were similarities in the music of the two composers, many of Ravel's piano pieces mirrored those of the classical period and not those of the impressionistic period.

In addition to the style of Debussy and Ravel, a composer by the name of *Igor Stravinsky,* illustrated complete change in the style of music. The rhythms in Stravinsky's music were very irregular. He challenged the ear by changing meters (time signatures) in the middle of a piece. Then he would return to the original meter leaving the listener wondering about the unexpected.

Another composer that reflected tension in his music was *Arnold Schoenberg.* Schoenberg was labeled a composer of *expressionism.* Like *impressionism,* expressionism is a term given to an art style. This style illustrates man in the modern

---

[8] Ibid. p.793.

world as described by twentieth century psychology, as "isolated, helpless to what he doesn't understand, prey to inner conflict, tension, anxiety and fear…"[9] Schoenberg's music suggests this with the use of much *dissonance* (melodic and harmonic tension), atonality (melodies and harmonies not based on one key center), which is the opposite of the traditional melodies and harmonies used in the Baroque, Classical and Romantic periods.

Composers:

- Claude Debussy
- Maurice Ravel
- Igor Stravinsky
- Arnold Schoenberg

[9] Ibid. p.853.

Quiz 7

1) What are the approximate dates of this period?

2) What was the historical background of the music of this period? Explain the
   a. thoughts of the people or any major events taking place that had a direct effect
   b. on the music during this time period.

3) What type of music flourished during this time period?

4) In what country(ies) did the music of this time flourish?

5) Who composed the music of this period?

6) Name some of the instruments used during this time period.

7) Listen to the music of this time period and describe what you hear. List
   a. some characteristics.

8) Listen to the music of this time period and explain what comes to your mind when you hear it.

9) Listen to the music of this time period and describe how it makes you feel when you hear it and explain what causes you to feel that way. For example, do you think it's the instruments, the tempo, the words or something else that you hear that causes you to feel the way that you do?

10) If you are in a classroom setting, listen to a particular song of this time period and create a scene without words that fits the music. Write it down then act out the scene. Have the class attempt to figure out the meaning of your scene. If you are studying music individually, listen to a particular song of this time period, create a story that fits the music. Record it in a journal.

Music Timeline Exercise

Directions: Draw a music timeline from the Ancient Period through the early Twentieth Century. (Remember time periods may overlap.) Include important key musicians and the instruments used in each period. Then once the timeline is completed, research and write a report on your favorite period in music history between the Ancient Period and the Twentieth Century. Include important dates, and instruments. Give a historical background of some of the key musicians and/or composers and explain why they had such an impact on the period they dominated.

Ancient Period -- Early Twentieth Century

African American Music

The Progression of Jazz, Some of Its "Unsung" Artists of Classical Music and the Progression of Popular Music from Gospel to Hip Hop

Lesson 21A

Jazz: Ragtime through Modern Jazz

Note: The following music selections can be used to reinforce the brief music history lessons given in this book, as well as serve as a guide to beginning a personal music collection. Each selection represents the music of the time period discussed. This is only a small taste of the wealth of music that exists. All of this music can be found in most major record stores.

Ragtime
Scott Joplin: *The Entertainer; Maple Leaf Rag*

Jelly Roll Morton
The Perfect Rag; Black Bottom Stomp

Blues

Vocal
Bessie Smith: *Down Hearted Blues; Back Water Blues; 'Taint Nobody's Business If I Do*

Instrumental
Sammy Price: *Honey Grove Blues*
Lionel Hampton: *Hamp's Boogie Woogie*

The Swing Era
Edward "Duke" Ellington: *Take the A Train*
William "Count" Basie: *One O'Clock Jump*
Benny Goodman: *Henderson Stomp*
Glenn Miller: *In the Mood*

Bop
Thelonious Monk: *Criss Cross; Blue Sphere*
Charlie Parker: *Yardbird Suite*
Dizzy Gillespie: *A Night in Tunisia*

Cool Jazz
Miles Davis: *Bag's Groove; Round Midnight*
Dave Brubeck: *The Way You Look Tonight; TimeOut*

Hard Bop
Horace Silver: *Six Pieces of Silver*

Free Jazz

Cecil Taylor: *Port of Call; Enter Evening*
John Coltrane: *Giant Steps*
Modern Jazz Piano Styles
Bill Evans: *Peace Piece; Waltz for Debby*
Herbie Hancock: *Maiden Voyage*
Chick Corea: *Matrix*
Keith Jarrett: *Forest Flower-Sunrise; The Celestial Hawk*

The Progression of Jazz

Jazz has a very diverse and rich history. It's an authentic American music created and developed by African Americans. Even though African Americans created and developed it, serious musicians of other races have contributed to it.

Studying the history of jazz can prove to be a very difficult task. It is not, nor has it ever been, music of clear distinction. To define jazz is to come up with a number of controversial definitions. For instance, some people define jazz as an improvised music that is "played first, and then written down later." This is an interesting concept but according to this definition, a Bach chorale (a form of music created by the German composer Johann Sebastian Bach who is discussed in the next chapter during the Baroque period) might be classified as jazz. It is not unlikely that Bach played his music before it was revised and documented. Others have said that music is considered jazz when it has a "preponderance of syncopated rhythm figures." [1] But doesn't some Spanish, classical, and rock music contain syncopation (the accenting of beats that normally aren't accented)? So, does that mean that these types of music are also classified as jazz? Still, others have combined the former two definitions and labeled jazz as both an improvised music that projects a jazz swing feeling. This definition is the more widely accepted one because it is broader than the previous two. However, this definition still is not completely satisfactory because it does not satisfy those who believe that jazz is also a music that uses "traditional jazz instruments (saxophones, trumpets, trombones etc.)"[2]

Because jazz is such a difficult term to define, a lot of different music is classified as such; mainly because of the perception people have about jazz. Because the definition of jazz is not etched in stone, people make jazz whatever they want it to be. This means that just about all African American music, including ragtime and blues are sometimes categorized as jazz. However, this perception is wrong. Jazz is a unique music, separate from ragtime and blues. But this is not to say that these other forms of African American music did not

---

[1] Mark C. Gridley, *Jazz Styles: History and Analysis* (Englewood Cliffs, New Jersey, 1991), p.6.
[2] Ibid. 7.

contribute to its development. As a matter of fact, these other styles of music had a lot to do with its development. That is why it is important to examine each of these other styles in a chronological fashion so that we get a clear picture of each characteristic style.

Ragtime Music (1890's–1920's)

Historical Background: Ragtime has a difficult history to trace. In early studies, it was believed that "ragtime" received its name because of its "ragged" rhythms. Other studies have associated the term with the rough and tattered clothing worn by some of the musicians developing this music. Yet, a third belief is that the word "rag" came from black clog dancing, derived from Irish clog dancing, which was called ragging. It consisted of syncopated rhythms produced by clapping, feet stamping and thigh slapping. However, its definite origin is unknown.

There are several different types of ragtime: *instrumental* (piano, banjo, and ragtime for bands), *vocal*, *syncopated waltzes*, and the *ragging of the classics* (European music). However, ragtime piano is the most commonly known.

About the Music: A ragtime piece is a composition in duple meter that centers on a basic left-hand pattern that "alternates from a single note or *octave* (two notes spaced eight notes apart), placed deep in the bass range of the piano, to a chord filling out the same harmony, that is placed in the mid-range of the piano. The notes are most often played on the strong beats of a 4/4 measure (beats 1 and 3) while the chords are played on the weak beats (beats 2 and 4)."[3] The bass line gives off an "oom-pah" sound as it accompanies a jagged syncopated melody.

Composers:

- **Scott Joplin** composed ragtime music that has European characteristics (characteristics that resembled Baroque music discussed in the next chapter)
- **James Scott** composed ragtime music that resembled Joplin's at first and then his melodies became shorter.
- **Joseph Lamb's** ragtime compositions were quite remarkable because he was a Caucasian man who was unfamiliar with African American culture and music but he was able to pick up this style from ragtime sheet music publications.

---

[3] Paul S. Machlin, *Stride: The Music of Fats Waller*, p.8.

Quiz 8

1) What are the approximate dates of this period?

2) What was the historical background of the music of this period? Explain the
    a. thoughts of the people or any major events taking place that had a direct effect
    b. on the music during this time period.

3) What type of music flourished during this time period?

4) In what country(ies) did the music of this time flourish?

5) Who composed the music of this period?

6) Name some of the instruments used during this time period.

7) Listen to the music of this time period and describe what you hear. List
    a. some characteristics.

8) Listen to the music of this time period and explain what comes to your mind when you hear it.

9) Listen to the music of this time period and describe how it makes you feel when you hear it and explain what causes you to feel that way. For example, do you think it's the instruments, the tempo, the words or something else that you hear that causes you to feel the way that you do?

10) If you are in a classroom setting, listen to a particular song of this time period and create a scene without words that fits the music. Write it down then act out the scene. Have the class attempt to figure out the meaning of your scene. If you are studying music individually, listen to a particular song of this time period, create a story that fits the music. Record it in a journal.

Jelly Roll Morton (1920's)

Historical Background: The unique piano style of Ferdinand Le Menthe, better known as Jelly Roll Morton, bridges the gap between the world of ragtime and early jazz.

Jelly Roll Morton (1890-1941) noted for his original style of music, has often been considered the link between ragtime and jazz, but he considered himself a jazz composer and pianist.

For about ten years Morton traveled, working mostly at music-related jobs. During this time of travel, Morton acquired a varied repertoire, including music by Scott Joplin and James Scott. He absorbed the styles of these composers, and incorporated them into his own. (His version of the *Maple Leaf Rag* is an excellent example of this.) In 1923, he continued his career as a recording soloist, bandleader, and composer. Three years later, he made a magnificent series of recordings with his band, the "Red Hot Peppers," in which composition and improvisation were carefully balanced. In 1929, he experienced some success as a composer when Melrose Brothers Music Company published about thirty of his compositions, including *London Blues*, *Perfect Rag*, *Black Bottom Stomp*, and *Wild Man Blues*.

About the Music: Most jazz historians consider Jelly Roll Morton, to be the first important jazz composer and pianist. He was capable of playing in both the ragtime and jazz styles, and was the first jazz musician to combine composition and *improvisation* (the act of making up music on the spot while playing; free-styling). He did this in an elaborate yet organized manner, making the end result a balance between the two.

What made his style of playing so exciting was his use of **eighth** notes (two eighth notes equal one quarter note (one beat)). He usually altered his eighth notes by playing them in a long-short pattern that produced a distinctive swing feel. He also used fewer **embellishments** (musical decorations) in his playing, making his pieces rhythmically buoyant and less characteristic of the ragtime style. Morton's piano playing simulated the sound of a jazz band by imitating such instruments as trumpets, clarinets and trombones and by using the stop-time solo break: "a technique used in jazz combos where the tempo stops when all group members except the soloist stop playing."[4] All of this, plus the fact that

---

[4] Mark C. Gridley, *Jazz Styles: History and Analysis*, p.17.

Morton incorporated a more blues-oriented New Orleans style into his rags, produced much activity and excitement in his music.

Instruments: Piano

Quiz 9

1) What are the approximate dates of this period?

2) What was the historical background of the music of this period? Explain the
    a. thoughts of the people or any major events taking place that had a direct effect
    b. on the music during this time period.

3) What type of music flourished during this time period?

4) In what country(ies) did the music of this time flourish?

5) Who composed the music of this period?

6) Name some of the instruments used during this time period.

7) Listen to the music of this time period and describe what you hear. List
    a. some characteristics.

8) Listen to the music of this time period and explain what comes to your mind when you hear it.

9) Listen to the music of this time period and describe how it makes you feel when you hear it and explain what causes you to feel that way. For example, do you think it's the instruments, the tempo, the words or something else that you hear that causes you to feel the way that you do?

10) If you are in a classroom setting, listen to a particular song of this time period and create a scene without words that fits the music. Write it down then act out the scene. Have the class attempt to figure out the meaning of your scene. If you are studying music individually, listen to a particular song of this time period, create a story that fits the music. Record it in a journal.

The Blues (1920's-1930's)

Historical Background: One of the important aspects of jazz, which distinguishes it from ragtime, is the presence of blues inflection. The **blues** (ca. 1900) evolved from several different African American musical utterances such as field hollers and cries, shouts, work songs, grunts and other sounds of expression that convey the deepest, innermost feelings of African Americans working under the conditions of slavery. It was originated with wandering minstrels, migrants, steel drivers, ditch diggers and other African Americans surviving an oppressed existence. After the development of the blues, these sounds were incorporated into the vocal music of African Americans and also imitated by jazz instrumentalists. For example, pianists would "bend notes" by striking two adjacent keys almost simultaneously, imitating **blue notes** (microtonally flattened third and seventh degrees of the diatonic scale), commonly used by the blues singers.

During the Great Migration (the movement of African Americans from the rural south to the northern urban areas), the blues began to develop into a new form and take on basic structure. It was referred to as "urban blues." This music contained chord progressions consisting of basic harmonies. It was developed in the cities and featured women singers such as Mamie Smith, Bessie Smith and Ma Rainey.

About the Music: The blues is more rhythmically fluid than ragtime with basic harmonies and repetitious melodic figures. It is unlike any other type of music except the **Negro Spiritual**. A spiritual is made up of the same melodies, harmonies and rhythms as the blues. As a matter of fact, the spiritual is the "sacred" counterpart to the blues. The only distinguishable difference between the two styles of music is the text. If the subject is not about God, the music is the blues. If the subject is about God, the music is a spiritual.

During the late 1920's and early 1930's, a new style of blues developed. It was a percussive style of piano blues called boogie-woogie. It consisted of blues chord progressions played over a forceful repetitive left-hand figure called a walking bass. But the progression of this music did not stop there. During the 1950's, boogie-woogie was mainly played by small bands, which evolved directly to the style of music known as rock and roll.

Instruments:

- Voice
- Piano
- Trumpet
- Saxophone
- Guitar Voice

Quiz 10

1) What are the approximate dates of this period?

2) What was the historical background of the music of this period? Explain the
   a. thoughts of the people or any major events taking place that had a direct effect
   b. on the music during this time period.

3) What type of music flourished during this time period?

4) In what country(ies) did the music of this time flourish?

5) Who composed the music of this period?

6) Name some of the instruments used during this time period.

7) Listen to the music of this time period and describe what you hear. List
   a. some characteristics.

8) Listen to the music of this time period and explain what comes to your mind when you hear it.

9) Listen to the music of this time period and describe how it makes you feel when you hear it and explain what causes you to feel that way. For example, do you think it's the instruments, the tempo, the words or something else that you hear that causes you to feel the way that you do?

10) If you are in a classroom setting, listen to a particular song of this time period and create a scene without words that fits the music. Write it down then act out the scene. Have the class attempt to figure out the meaning of your scene. If you are studying music individually, listen to a particular song of this time period, create a story that fits the music. Record it in a journal.

The Swing Era (Late 1920's-1940's)

Historical Background: It was amazing that the **Swing** era flourished so well since the thirties were marked by a worldwide depression that hurt the record industry tremendously. However, ironically enough, the depression kept music lovers at home listening to live bands on the radio. This launched this explosive musical period within the next few years.

About the Music: The Swing era received its name because of the "long-short" eighth note rhythmic pattern used. This pattern caused the music to jump at lively speeds. This was a looser and freer style of playing first introduced in the music of Jelly Roll Morton.

This period was referred to as the **Big Band** era also. It acquired this name because the bands that played during this time were made up often of ten or more musicians.

Instruments: (Three Sections)

<u>Brass Section</u>:
Trumpets
Trombones

<u>Woodwind Section</u>:
Clarinet
Alto Saxophone
Tenor Saxophone
Baritone Saxophone

Note: Saxophones are made of brass but the instruments they originated from were wooden and since they have reeds and are played in the manner of traditional wooden instruments they are considered woodwind instruments.

<u>Rhythm Section</u>:
Piano
Guitar
Bass
Drums

Composers/Bandleaders/Musicians:

- Edward "Duke" Ellington (pianist, bandleader)
- William "Count" Basie (pianist, bandleader)
- Benny Goodman (clarinet)
- Glenn Miller (trombone, bandleader)

Quiz 11

1) What are the approximate dates of this period?

2) What was the historical background of the music of this period? Explain the
    a. thoughts of the people or any major events taking place that had a direct effect
    b. on the music during this time period.

3) What type of music flourished during this time period?

4) In what country(ies) did the music of this time flourish?

5) Who composed the music of this period?

6) Name some of the instruments used during this time period.

7) Listen to the music of this time period and describe what you hear. List
    a. some characteristics.

8) Listen to the music of this time period and explain what comes to your mind when you hear it.

9) Listen to the music of this time period and describe how it makes you feel when you hear it and explain what causes you to feel that way. For example, do you think it's the instruments, the tempo, the words or something else that you hear that causes you to feel the way that you do?

10) If you are in a classroom setting, listen to a particular song of this time period and create a scene without words that fits the music. Write it down then act out the scene. Have the class attempt to figure out the meaning of your scene. If you are studying music individually, listen to a particular song of this time period, create a story that fits the music. Record it in a journal.

Bop (Early 1940's – Mid 1950's)

Historical Background: The swing era produced a number of styles that flowed smoothly into modern jazz styles. **Bop**, which is considered the first **modern** jazz style, developed from the early 1940's to the middle 1950's. The music prior to the 1940's was considered *classic* jazz. One of the traits that distinguish bop from swing is smaller combos (bands).
About the Music: Bop musicians took *swing* music and added their own techniques. They even studied contemporary European music by composers like Stravinsky.
The melodies were more complex in bop than in swing and the rhythms were more varied and less predictable. Because of this, it was more difficult to dance to it therefore, bop wasn't as popular as swing.

Musicians:

- Charlie Parker (alto saxophone)
- Dizzy Gillespie (trumpet)
- Miles Davis (trumpet)
- Thelonious Monk (piano)
- Earl "Bud" Powell (piano)

Cool Jazz (1950's)

Historical Background: The term, **cool jazz**, was given to jazz musicians of this time by journalists and fans. The musicians did not accept it because it seemed to suggest that their music had a lack of passion when in fact, it was used to describe music with soft sounds.
About the Music: Cool jazz is a simpler, softer type of music that contains more melody and lighter tone qualities than bop. On the other hand, there is some music classified as cool jazz that has similar qualities as bop.

Musicians:

- Miles Davis (trumpet)
- Stan Getz (saxophone)
- Dave Brubeck (piano, composer, arranger)
- Lennie Tristano (piano)

Quiz 12

1) What are the approximate dates of this period?

2) What was the historical background of the music of this period? Explain the
   a. thoughts of the people or any major events taking place that had a direct effect
   b. on the music during this time period.

3) What type of music flourished during this time period?

4) In what country(ies) did the music of this time flourish?

5) Who composed the music of this period?

6) Name some of the instruments used during this time period.

7) Listen to the music of this time period and describe what you hear. List
   a. some characteristics.

8) Listen to the music of this time period and explain what comes to your mind when you hear it.

9) Listen to the music of this time period and describe how it makes you feel when you hear it and explain what causes you to feel that way. For example, do you think it's the instruments, the tempo, the words or something else that you hear that causes you to feel the way that you do?

10) If you are in a classroom setting, listen to a particular song of this time period and create a scene without words that fits the music. Write it down then act out the scene. Have the class attempt to figure out the meaning of your scene. If you are studying music individually, listen to a particular song of this time period, create a story that fits the music. Record it in a journal.

Hard Bop (1950's-1960's)

Historical Background: The term **hard bop**, like that of cool jazz was given to the style of jazz, as described by journalists, to categorize a musician's style of playing. The musicians didn't approve of the label.

About the Music: Hard Bop is similar to bop but it has simpler improvised lines and an emphasis on consistent swinging. Horace Silver (b. 1928) is its most significant pianist.

Silver was mostly known for his work as a composer and bandleader, but his piano style showed originality and depth. His melodic phrases tend to lean toward the shorter style of writing. He turned the longer, moving passages of early bop into shorter, substantiated phrases. But his main concern for his playing was to produce clear and distinct melodies, not the rapid flowing melodies Lennie Tristano tried to master.

Musicians:

- Horace Silver (piano)
- Clifford Brown (trumpet)
- Art Blakey (drums)

Free Jazz *(1950's)*

Historical Background: Free jazz was the most radically new development in jazz in the 1950's. The term free jazz was given to describe jazz that features "free" improvisation outside of "preset" jazz forms.

About the Music: One of its principal figures was the pianist, Cecil Taylor (b. 1929). As a pianist, composer, and bandleader, Taylor developed a controversial style of jazz during the late '50's and early '60's that sometimes produced atonality (music not played in a specific key), striking dissonances and a sense of chaos.

Taylor's piano style was somewhat like his ensemble style. He focused more on the texture of his music than on singable melodies. His notes tended to be layered on top of one another, creating a rich and explosive sound. He also played these notes at incredible speeds and in a percussive manner. His music did consist of rhythmic syncopation (rhythms that contain accents on beats that are not normally accented), but not to the point of swinging, which means that it lacked the buoyancy of "traditional" jazz. Taylor's style was often critically acclaimed, but did not become generally popular.

Musicians:

- Cecil Taylor (piano)
- John Coltrane (tenor and soprano saxophone)

Quiz 13

1) What are the approximate dates of this period?

2) What was the historical background of the music of this period? Explain the
   a. thoughts of the people or any major events taking place that had a direct effect
   b. on the music during this time period.

3) What type of music flourished during this time period?

4) In what country(ies) did the music of this time flourish?

5) Who composed the music of this period?

6) Name some of the instruments used during this time period.

7) Listen to the music of this time period and describe what you hear. List
   a. some characteristics.

8) Listen to the music of this time period and explain what comes to your mind when you hear it.

9) Listen to the music of this time period and describe how it makes you feel when you hear it and explain what causes you to feel that way. For example, do you think it's the instruments, the tempo, the words or something else that you hear that causes you to feel the way that you do?

10) If you are in a classroom setting, listen to a particular song of this time period and create a scene without words that fits the music. Write it down then act out the scene. Have the class attempt to figure out the meaning of your scene. If you are studying music individually, listen to a particular song of this time period, create a story that fits the music. Record it in a journal.

Modern Jazz Piano Styles

Bill Evans

Historical Background: Since the 1960's, pianists have continued to play an important role in the history of jazz. Bill Evans, Herbie Hancock, Chick Corea, and Keith Jarrett have been very influential figures. Although their styles may be easily distinguished from each other, all four were influenced by non-jazz musical traditions.

Bill Evans (1929-1980) studied piano at Southeastern Louisiana University but he did not begin his career as a jazz pianist until after he served in the army. In 1956, Evans made his first recording with his own group. Evans derived his style from such musicians as Lennie Tristano, from whom he developed his long, flowing melodies. Bits of Earl Bud Powell are heard and Evans' blues-like figures are thought to be derived from Horace Silver.

After the year 1959, Evans gave his left hand a more important role. He began to sustain chords in his left hand rather than just touch upon them. He was much interested in the harmonies of Debussy and Ravel which gave his music an *impressionistic* quality (see Music during the *Early to Mid-Twentieth Century*: 1900-1951 in the European music history section of this book).

About the Music: The music of Bill Evans had a very memorable sound. He used delicate melodies to capture his audiences. Some have even referred to his sound as harp-like. Evans was capable of taking a small musical idiom and expanding its melodies, harmonies and rhythms. He rarely used a percussive approach to his music but he did like to build tension that did not always resolve.

Rhythmically, Evans' music was very involved. He would phrase musical passages across bar lines, eliminating a definite pulse or beat. He would also cloud the meter (time signature) by accenting the upbeats of measures instead of the downbeats. He would run his long, quick and smooth melodies across several measures and the downbeat would never become evident.

Herbie Hancock

Historical Background: Herbie Hancock (b. 1940) was born in Chicago and began studying the piano at the age of seven. When he was eleven, he performed the first movement of a Mozart concerto (see Music during the *Classical Period* in the European music history section of this book for the definition of a concerto) in a young people's concert. While he was in high school, Hancock formed his own jazz ensemble. In 1960, when he graduated from Grinnell College, he was performing with Coleman Hawkins and Donald Byrd in Chicago jazz clubs. In 1963, Hancock joined Miles Davis's quintet where his style of music became evident.

About the Music: A great contribution Hancock brought to jazz was how to put more variety into the left hand. He, too, used *impressionistic* characteristics (see Music in the *Early to Mid Twentieth Century: 1900-1951*) in the European music history section of this book) by adopting some of Debussy's and Ravel's ideas including polyrhythmic (the use of more than one rhythm) approaches to accompaniment. He also based his improvisations on modes (see Music During the *Ancient Period* in the European music history section of this book for clarification of modes) rather than chord changing to give his sound more color. This affected the mood of his music. He did not want to be confined to just following traditional chord progressions, so along with using modes, he would improvise melodies and harmonies to obtain a certain mood, and when he had the urge, he would instantly change his melodies and harmonies to create another one.

Chick Corea

Historical Background: The music history of Chick Corea (b. 1941) began when he studied the piano at the age of four. He developed an interest in jazz after listening to recordings of Dizzy Gillespie, Charlie Parker and Billy Eckstine (jazz vocalist during the forties). Horace Silver and Earl Bud Powell also influenced him. Corea's first professional performances were in the Latin bands of Mongo Santamaria and Willie Bobo in 1962. Then in 1964, he worked with Blue Mitchell and recorded his own compositions for Blue Note records.

About the Music: Corea's style originated from the style of Bud Powell, Horace Silver, Bill Evans, McCoy Tyner (jazz pianist during the sixties) and it also reflects characteristics of twentieth century European composers Paul Hindemith and Bela Bartok. Corea used quartal harmony (chords voiced in fourths) to give his music an unusual quality. He also used the pentatonic scale (major scale without the fourth and seventh scale degrees) more often than other contemporary jazz musicians.

Some of Corea's music was influenced by the sounds of Latin America. The rhythms took on a double-time feeling with very little emphasis on any particular beat. His percussive approach and uplifting style of comping added to the Latin American feel.

Corea had a talent for adding variety to his music. He would sometimes have different sections in a piece of music with completely different rhythms and harmonies in each. Some of these pieces portrayed the Spanish flavor mentioned earlier along with rock-solid bass lines to accompany it. It was particularly this style of music that made him popular as a jazz-rock fusion pianist.

Keith Jarrett

Historical Background: Keith Jarrett (b. 1945) began studying the piano at the age of three. By the age of seven he was composing and improvising music. In 1962, he studied at the Berklee College of Music in Boston. He moved to New York in 1965 and later joined Blakey's Jazz Messengers. Jarrett received recognition after joining Charles Lloyd's quartet in 1966. This group was one of the first to explore improvisational styles extensively. In 1969, Jarrett continued his success as an accompanist for trumpeter Miles Davis. Jarrett, slightly younger than both Hancock and Corea, proved to be at least as talented and perhaps even more so. Jarrett's musical imagination appeared to be vast.

About the Music: He came up with the most original ideas and put them together as if each idea were a piece to a musical puzzle. Jarrett derived his style from Bill Evans, Ornette Coleman (alto saxophonist), and European composers Bela Bartok and Maurice Ravel. It also reflected American gospel and country music. He could take musical ideas from each of those styles and blend them together, creating a rich and diverse sound. Some of his melodies resemble long saxophone lines. Like Coleman, he would maintain legato (smooth, connected) lines and sometimes create music that lacked a definite tonal center.

Quiz 14

1) What are the approximate dates of this period?

2) What was the historical background of the music of this period? Explain the
   a. thoughts of the people or any major events taking place that had a direct effect
   b. on the music during this time period.

3) What type of music flourished during this time period?

4) In what country(ies) did the music of this time flourish?

5) Who composed the music of this period?

6) Name some of the instruments used during this time period.

7) Listen to the music of this time period and describe what you hear. List
   a. some characteristics.

8) Listen to the music of this time period and explain what comes to your mind when you hear it.

9) Listen to the music of this time period and describe how it makes you feel when you hear it and explain what causes you to feel that way. For example, do you think it's the instruments, the tempo, the words or something else that you hear that causes you to feel the way that you do?

10) If you are in a classroom setting, listen to a particular song of this time period and create a scene without words that fits the music. Write it down then act out the scene. Have the class attempt to figure out the meaning of your scene. If you are studying music individually, listen to a particular song of this time period, create a story that fits the music. Record it in a journal.

Lesson 21B

African American Music: Some of its "Unsung" Artists in Classical Music

Classical Music by African Americans

William Grant Still: *Afro-American Symphony*
Florence Price: *Symphony in E Minor*
Paul Robeson: recordings from the plays *Show Boat*; *Porgy and Bess* and *Othello*
Marion Anderson: recordings from Verdi's opera, *Un Ballo in Maschera*
William C. Warfield: recordings from Gershwin's *Porgy and Bess*
Leontyne Price: recordings from Verdi's opera, *Il Trovatore*; Puccini's production of *The Girl of the Golden West* and Samuel Barber's production of *Antony and Cleopatra*
George Shirley: recordings from Strauss's opera, *Die Fledermaus* and Puccini's *La Boheme*
Jessye Norman: recordings from Wagner's opera, *Tannhauser*; and Strauss's *Ariadne auf Naxos*
Later recordings include Berg's *Lulu Suite*; *Berlioz's Les nuits d'ete* and *Romeo and Juliet*; Bizet's *Carmen* and an album entitled *Lucky to be Me*
Kathleen Battle: recordings from Wagner's opera; *Tannhauser*; Strauss's *Ariadne auf Naxos* and Mahler's *Symphony No. 4*
Scott Joplin: *Treemonisha*

African-Americans in the World of Classical Music (1850-)

Historical Background: Music created by African Americans always had a place in American history but most of the time it was not recognized. Like mentioned in the previous lesson, jazz is America's first authentic music and it has left a mark on American history that never can be erased. However, at the same time jazz was developing, there were African American artists leaving their stamp in the world of European music.

In the nineteenth century, slaves who could play instruments such as the piano, flute or violin entertained their masters at dinner parties and balls. However, because of their talent, some of these same slaves went on to perform European music in their own concerts and continued performing as professional musicians after they were freed from slavery. One such musician was a blind pianist by the name of Thomas Green Bethune

(1849-1909) who demonstrated his musical genius by performing nearly 7000 compositions by famous European composers such as Bach, Beethoven, Chopin, Liszt and others in Europe as well as in the United States.

In the twentieth century, an important composer by the name of William Grant Still (1895-1978) composed *Afro-American Symphony*, the first symphonic work written by an African American composer that was performed by a major symphony orchestra (The Rochester Philharmonic Symphony) in 1931.[8]

In 1933, Florence Price (1888-1953) was the first African American female to have her composition, *Symphony in E Minor*, performed by the famous Chicago Symphony Orchestra.[9]

Other African American artists that mastered the European style of music are vocalists, Harry Burleigh (1866-1949), Paul Robeson (1898-1976), Marion Anderson (1902-1993) who gave a most memorable concert on the steps of the Lincoln Memorial after being barred from singing at Constitutional Hall by the Daughters of the American Revolution (D.A.R.) because she was African-American, William C. Warfield (1920-), Leontyne Price (1927-) who opened a convention of the Daughters of the American Revolution (D.A.R.) in Constitutional Hall in honor of Marion Anderson who was barred from singing there forty-three years earlier, George Shirley (1934-), Jessye Norman (1945-) and Kathleen Battle (1948-).

Some of the extra-ordinary African American instrumentalists that penetrated the "classical" world of music include, pianist/composer Scott Joplin (1868-1917) who is best known for composing ragtime music but has composed a ballet and two operas (one of which the music was lost and the other being the famous *Treemonisha*), cellist/composer Kermit Moore (1929-), bassoonist/conductor Karl Hampton Porter (1939-), and pianist Andre' Watts (1946-) who performed masterfully under the well-known conductor, Leonard Bernstein of the New York Philharmonic.

About the Music: Classical music by African American composers fell into two categories: "black-stream" music named after Gunther Schuller's *Third Stream* and traditional European music. Black-stream music is classical music that conveys evidence

---

[8] Kenneth Estell, *African America: Portrait of a People*, p. 428.
[9] Ibid. p. 429.

of the composer's ethnic background. The traditional European style of music shows no evidence of the composer's ethnic background. Whichever the style, this music composed by African Americans is important and is, without question, worthy of serious study.

Quiz 15

1) What are the approximate dates of this period?

2) What was the historical background of the music of this period? Explain the
   a. thoughts of the people or any major events taking place that had a direct effect
   b. on the music during this time period.

3) What type of music flourished during this time period?

4) In what country(ies) did the music of this time flourish?

5) Who composed the music of this period?

6) Name some of the instruments used during this time period.

7) Listen to the music of this time period and describe what you hear. List
   a. some characteristics.

8) Listen to the music of this time period and explain what comes to your mind when you hear it.

9) Listen to the music of this time period and describe how it makes you feel when you hear it and explain what causes you to feel that way. For example, do you think it's the instruments, the tempo, the words or something else that you hear that causes you to feel the way that you do?

10) If you are in a classroom setting, listen to a particular song of this time period and create a scene without words that fits the music. Write it down then act out the scene. Have the class attempt to figure out the meaning of your scene. If you are studying music individually, listen to a particular song of this time period, create a story that fits the music. Record it in a journal.

Lesson 21C

African American Music: The Progression of Popular Music from Gospel to Hip Hop

Because African American musicians made such an impact on the origin of American music, it is impossible to overlook the influence they had and continue to have on the progression of modern jazz and today's popular music. It was already mentioned in lesson 16 how the music of contemporary jazz artists reflected both, earlier forms of jazz as well as characteristics of European music. It is no different for artists of popular music of yesterday or today. Popular music is deeply rooted in the sounds of gospel, rhythm and blues, rock and roll, soul, funk and disco music.

Gospel
Thomas A. Dorsey: *Precious Lord*; *There'll Be Peace in the Valley* and *The Lord Will Make a Way*
Mahalia Jackson: *In the Upper Room*; *Move on up a Little* and *Let the Holy Ghost Fall Down on Me*

Rhythm and Blues
Ray Charles: *I Got a Woman*; *What I'd Say*
Sam Cooke: *You Send Me*; *Bring It on Home to Me* and *Little Red Rooster*

Rock and Roll
Little Richard: *Tutti Frutti*; *Long Tall Sally*
Chuck Berry: *Maybelline*; *Roll Over Beethoven*; *Oh Baby Doll*; and *Johnny B. Goode*
Ike and Tina Turner: *Poor Fool*; *Proud Mary* and *Shake a Tail Feather*
Fats Domino: *Blueberry Hill*; *Whole Lotta Lovin'; Ain't That a Shame* and *I'm Walkin'*

Soul
Aretha Franklin: *I Never Loved a Man (The WayTthat I Loved You); Do Right Woman, Do Right Man* and *Respect*
James Brown: *I Feel Good*; *Please, Please, Please* and *I'm Black and I'm Proud*
Curtis Mayfield: *People Get Ready*, *Choice of Colors*, *Superfly*

Motown
The Marvelettes: *Please Mister Postman*
Marvin Gaye: *Pride and Joy* and *Heard It Through the Grapevine*
Smokey Robinson and the Miracles: *Tears of a Clown*; *Tracks of My Tears* and *Shop Around*
The Supremes: *Baby Love* and *Can't Hurry Love*
The Temptations: *My Girl* and *Ain't Too Proud to Beg* and *Papa Was a Rollin' Stone*

Stevie Wonder: *Singed Sealed and Delivered (I'm Yours)*; *Uptight (Everything's Alright)* and *For Once in My Life*
Martha and the Vandellas: *Love Is Like a Heatwave*
The Four Tops: *Sugar Pie, Honey Bunch*
The Jackson Five: *I Want You Back*; *ABC*; *I'll Be There*; *The Love You Save* and *Who's Lovin'* You
Gladys Knight and the Pips: *I Heard It Through the Grapevine* and *Neither One of Us (Wants to Be the First to Say Goodbye)*

Funk
Earth, Wind and Fire: *Boogie Wonderland*; <u>Dancing in September</u> and *Reason*s
Rufus and Chaka Khan: *Tell Me Somethin' Good*
Parliament: *Atomic Dog*; *Give up the Funk* and *Flashlight*
Kool and the Gang: *Celebration*
The Ohio Players: *Fire* and *Skin Tight*
The Commodores: *Brick House*

Disco
Donna Summer: *On the Radio*; *Last Dance*
Thelma Houston: *Don't Leave Me This Way*
Peaches and Herb: *Shake Your Groove Thing* and *Reunited*
Gloria Gaynor: *I Will Survive*
Isaac Hayes: *Disco Connection*
The Trammps: *Disco Inferno*

Rhythm and Blues in the 1970's
Al Green: *For the Good Times* and *Let's Stay Together*
Barry White: *Can't Get Enough of Your Love, Babe* and *What Am I Gonna Do with You*
Roberta Flack: *The First Time Ever I Saw Your Face*; *Killing Me Softly (with His Song)* and *The Closer I Get to You* (Sung with Donny Hathaway)
Donny Hathaway: *The Closer I Get to You* (sung with Roberta Flack) and *This Christmas*
Harold Melvin and the Blue Notes: *If you Don't Know Me by Now*; *The Love I Lost* and *Wake Up Everybody*
The Fifth Dimension: *You Don't Have to be a Star to Be in My Show*
The Isley Brothers: *Secret Lover*

Rhythm and Blues in the 1980's
Luther Vandross: *Superstar*; *So Amazing*; *Stop to Love* and *A House Is Not a Home*
Anita Baker: *Sweet Love*
Freddie Jackson: *Rock with Me for Old Time Sake*

Pop in the 1980's and 1990's
Michael Jackson: *Rock With You*; *Off the Wall*; *Beat It*; *Billie Jean* and *Thriller*
New Edition: *Candy girl*; *Is This the End*; *Popcorn Love* and *She's Gives Me a Bang*
Lionel Richie: *All Night Long*; *Dancing on the Ceiling* and *Say You, Say Me*
Klymaxx: *Meeting in the Ladies Room*; *I Miss You* and *The Men All Pause*

Lisa Lisa and the Cult Jam: *I Wonder If I Take You Home*; *Head to Toe* and *All Cried Out*
DeBarge: *Rhythm of the Night*; *Time Will Reveal* and *Love Me in a Special Way*
Stephanie Mills: *Something in the Way You Make Me Feel*; *I Feel Good* and *Comfort of a Man*
Cameo: *Word Up*
Ready for the World: *Oh, Sheila*
Janet Jackson: *Control*; *What Have You Done for Me Lately* and *Rhythmnation*
Whitney Houston: *Saving All My Love for You* and You *Give Good Love*
Mariah Carey: *Vision of Love*; *Love Takes Time*
Boys II Men: *Motown Philly* and *End of the Road*

Rap
The Sugar Hill Gang: *Rapper's Delight*
Run DMC: *My Adidas*; and *It's Like That*
The Furious Five: *The Message*
M.C. Hammer: *Can't Touch This*
Ice-T: *I Must Stand*
Public Enemy: *Fight the Power*; *Fear of a Black Planet*
Ice Cube: *It's a Good Day*
Tupac Shakur: *Keep Your Head Up* and *Thug Mansion* (explicit language)
Notorious BIG: *Juicy* (explicit language)
Dougie Fresh: *The Show*
Fat Boys: *Fat Boys*
Dr. Dre and Snoop Dogg: *Ain't Nothin' but a 'G' Thing*

Hip Hop
Queen Latifah: *U-N-I-T-Y*
Mary J. Blige: *Reminisce*; *No More Drama*
P. Diddy: *I'll Be Missing You*
Destiny's Child: *I'm a Survivor*
Nelly: *Country Grammar* and *EI*
Angie Stone: *No More Rain*

Gospel Music (1850-1970)

Historical Background: Gospel music has played and continues to play a major role in the African American experience. Ever since the beginning of slavery, slaves sang about God and His plan for delivering them out of bondage as a means of hope, even escape from their oppressed existence.

About the Music: The syncopated African rhythms imbedded in the souls of the slaves combined with the European melodies of their masters brought about a new style of music. This music (known today as the **Negro Spiritual***)* incorporated blues inflections that imitated the shouts, grunts, groans, and other sounds of expression of a downtrodden people. It laid the foundation for gospel music. By the turn of the century, gospel music became very popular because even though African Americans were free from slavery, they weren't free from the burdens of racism in America. Poverty and economic struggles were a way of life in the African American community.

Thomas A. Dorsey, a jazz and blues pianist who became a **Spiritual** songwriter in the early 1900's inspired hope to African Americans with such songs as *Precious Lord, There'll Be Peace in the Valley* and *The Lord Will Make a Way*. His blues/jazz style of religious music swept the nation in the 1930's when it was first performed at the National Baptist Convention. Because of the popularity of gospel music, the recording industry began recording this music as "race records" (a title given to African American recordings)[10] and it continued to flourish and then reach outside of the African American community.

Another artist that had a major impact on gospel music was Mahalia Jackson, known as the "world's greatest gospel singer."[11] She toured Europe and recorded her gospel hit, *In the Upper Room*. Other well-known pieces were *Move on up a Little*, and *Let the Holy Ghost Fall on Me* which won the French Academy's Grand Prix du Disque award. Because Jackson was the best in her field, her voice graced the White House, London's Albert Hall, the 1963 March on Washington and Martin Luther King, Jr.'s funeral ceremony in 1968. Her music is still an inspiration today.

---

[10] Ibid. p. 539.
[11] Ibid. p. 569.

Rhythm and Blues (1940-1955)

Historical Background: During the mid-1940's while gospel music was making its mark in the music industry, non-religious music, possessing similar characteristics to those of gospel music, was making history as well. Like that of gospel music, this secular (non-religious) music was very rhythmic and consisted of blues, boogie-woogie and swing jazz thus making it almost indistinguishable from gospel music. This was the birth of **Rhythm and Blues**.

About the Music: The music of artists, Ray Charles and Sam Cooke are perfect examples of this. Ray Charles' hit, *I Got a Woman*, is based on the gospel song, *My Jesus Is All the World to Me* and his song, *What I'd Say* has the call-and-response pattern used when singing spirituals in the early African American churches.[12]

Sam Cooke, too, crossed over from gospel music to rhythm and blues. His soft, gospel-style of singing is demonstrated in his hit, *You Send Me*, which made him one of the most popular singers in rhythm and blues in the late 1950's.

Rock and Roll (1955-1965)

Historical Background: Rhythm and blues remained popular throughout the 1940's and early 1950's, because during this time, the style of jazz, with the birth of **bop, cool jazz, hard bop** and **free jazz** (see lesson 17), was no longer considered dance music. Therefore, in order to fulfill this musical void, rhythm and blues picked up where **swing** music (see lesson 17) of the 1930's left off as dance music. However, in the mid 1950's, rhythm and blues became swept under a new title. This title was **Rock and Roll**. The term rock and roll was used, first, by Caucasians in the media to describe music by Caucasian musicians at this time. However, because rhythm and blues became so popular, the same media used this term to describe the music of both, Caucasians and African Americans which in turn took the racial "sting" out of the music created by African Americans. This gave rhythm and blues artists the opportunity to perform for multi-racial audiences nationwide. Therefore, the music known as rock and roll by artists

---

[12] Ibid. p. 542.

such as Little Richard, Chuck Berry, Ike and Tina Turner and "Fats" Domino is actually rhythm and blues under a more "socially acceptable" title.

About the Music: Rock and roll music was played by small bands that consisted of a piano, guitars and drums. Generally, it had a fast tempo (rate of speed) and can be identified by its repetitive bass line making it a descendant of the early jazz style called **boogie-woogie** (see lesson 17). Boogie-woogie is best described as blues chords played forcefully over a repetitive left hand on the piano.

Quiz 16

1. What are the approximate dates of this period?

2. What was the historical background of the music of this period? Explain the
    a. thoughts of the people or any major events taking place that had a direct effect
    b. on the music during this time period.

3. What type of music flourished during this time period?

4. In what country(ies) did the music of this time flourish?

5. Who composed the music of this period?

6. Name some of the instruments used during this time period.

7. Listen to the music of this time period and describe what you hear. List
    a. some characteristics.

8. Listen to the music of this time period and explain what comes to your mind when you hear it.

9. Listen to the music of this time period and describe how it makes you feel when you hear it and explain what causes you to feel that way. For example, do you think it's the instruments, the tempo, the words or something else that you hear that causes you to feel the way that you do?

10. If you are in a classroom setting, listen to a particular song of this time period and create a scene without words that fits the music. Write it down then act out the scene. Have the class attempt to figure out the meaning of your scene. If you are studying music individually, listen to a particular song of this time period, create a story that fits the music. Record it in a journal.

Soul (1960-1975)

Historical Background: As time progressed, a new style of music called **Soul** became a powerful force in the African American community. Soul, with its gospel influence, dominated the 1960's and had a major role in the Civil Rights Movement.

About the Music: Artists such as Aretha Franklin, Queen of Soul and James Brown, the Godfather of Soul defined this music. Both artists used the rich gospel sound in their music demonstrated by rhythmic shouts and screams and pulsating rhythms that drove the music. James Brown's *I'm Black and I'm Proud* became an anthem during this time.

Curtis Mayfield, another soul Veteran became popular with his music wrapped around social issues affecting the African American community. His songs, *People Get Ready* and *Choice of Colors* are excellent examples of this. Also, he composed the music for the 1972 hit film, *Superfly*, a movie depicting a drug-dealer wanting to leave the street life run by dishonest law enforcement.

The Motown Influence (1959-1984)

Historical Background: Because the soul sound was so popular, record labels capitalized on it. In 1959, one label in particular, **Motown Records**, located in Detroit Michigan, became the music machine that took this music to another level and changed the course of popular music forever. Its founder, Berry Gordy, employed Detroit's best jazz and classical musicians to accompany his singers and hired excellent choreographers to produce smooth and refined acts that attracted both the African American and Caucasian middle-class communities. Such acts produced by Motown were the Marvelettes, Marvin Gaye, Smokey Robinson and the Miracles, the Supremes, the Temptations, Stevie Wonder, Martha and the Vandellas, the Four Tops and of course, the Jackson Five.

About the Music: Without this successful African American music business enterprise called Motown, the importance of vocal harmony would not have been emphasized, the art of "doo-wop" never would have flourished and the smooth sultry sound of rhythm and blues would not have had such an impact on the music industry.

Quiz 17

1. What are the approximate dates of this period?

2. What was the historical background of the music of this period? Explain the thoughts of the people or any major events taking place that had a direct effect on the music during this time period.

3. What type of music flourished during this time period?

4. In what country(ies) did the music of this time flourish?

5. Who composed the music of this period?

6. Name some of the instruments used during this time period.

7. Listen to the music of this time period and describe what you hear. List some characteristics.

8. Listen to the music of this time period and explain what comes to your mind when you hear it.

9. Listen to the music of this time period and describe how it makes you feel when you hear it and explain what causes you to feel that way. For example, do you think it's the instruments, the tempo, the words or something else that you hear that causes you to feel the way that you do?

10. If you are in a classroom setting, listen to a particular song of this time period and create a scene **without words** that fits the music. Write it down then act out the scene. Have the class attempt to figure out the meaning of your scene. If you are studying music individually, listen to a particular song of this time period, create a story that fits the music. Record it in a journal.

Funk (1970's)

Historical Background/About the Music: During the early 1970's, still, while Motown was producing a more refined style of music with softer qualities, music that possessed a harder edge with its driving rhythms and the introduction of electric instruments emerged. This music was given the name **Funk**. Rhythmically, funk music can be traced back to soul music but because of the use of electric guitars, basses and synthesizers and then later horn sections, this music took on its own life form. Jimi Hendrix was instrumental in developing this music with his innovative driving electric guitar sound. Other major funk artists are Earth, Wind and Fire, Rufus and Chaka Khan, Parliament and later Kool and the Gang, the Ohio Players and the Commodores.

Disco (1970's)

Historical Background/About the Music: At the same time Funk was driving music one way, a lighter form of dance music called **Disco** was hitting dance clubs hard with its fast-moving tempo and sultry lyrics. Even though disco was short-lived, it featured major artists such as Donna Summer, Thelma Houston, Peaches and Herb, Gloria Gaynor, Isaac Hayes and the Trammps.

Rhythm and Blues of the 1970's Meet "Pop" in the 1980's and 1990's

Historical Background/About the Music: During the 1970's rhythm and blues portrayed yet another style; a slower style; a mellow style with the sounds of Al Green, Barry White, Roberta Flack, Donny Hathaway, Patti LaBelle, whose career not only began in the 1950's with an all-female group but continues to flourish today, Harold Melvin and the Blue Notes, The Fifth Dimension the Isley Brothers and other important artists. This mellow style continued into the 1980's with the sounds of the most sophisticated love ballads by Luther Vandross, Anita Baker and Freddie Jackson, who's style is "classic" today.

While the music of rhythm and blues was mellowing out in the 1980's, at the same time a more energetic music labeled by the media as **Pop** appeared on the radio waves. Michael Jackson (formerly of the Jackson Five) reigned and continues to reign as the "King of Pop."

Other Pop artists during this time were New Edition whose harmonies closely resembled that of the Jackson Five, Lionel Richie (formerly of the Commodores) who before this time in the late 1970's dominated the music charts with beautiful love ballads, a female band called Klymaxx, Lisa Lisa and the Cult Jam, DeBarge, Stephanie Mills, Cameo, and Ready for the World and countless others. The music of Rick James and Prince flourished during this time also but their style resembled more of the funk style than the rhythm and blues/pop style because the driving electronic guitar dominated their sound. Also, as in the Funk era, they used other electronic equipment such as synthesizers, which exploded during this time period. Herbie Hancock, the jazz pianist mentioned in the previous lesson, electrified the music world with his innovative use of synthesizers and rhythm machines demonstrated in his composition, *Rock It*. He brought electronic music to new heights.

However, the music industry never would have reached the level that it did during this time without the introduction of visual effects to accompany the music of this time. This led to the dynamic world of music videos that became the new crave of the 1980's that continues today. With this new wave of creativity, now music can be "seen" as well as heard.

Later during the 1980's, young Miss Janet Jackson (the youngest sister of Michael Jackson) was making her way in the music world along with the rich sounds of Whitney Houston who clearly shows evidence of a Gospel background in her style of singing. Both of these ladies continued to make hit records well into the 1990's and beyond. Other important artists of the 1990's were Mariah Carey, Boys II Men who brought back the rich vocal harmonies that dominated the Motown Sound, Guy, Keith Sweat, Al B. Sure and a host of others.

Quiz 18

1. What are the approximate dates of this period?

2. What was the historical background of the music of this period? Explain the thoughts of the people or any major events taking place that had a direct effect on the music during this time period.

3. What type of music flourished during this time period?

4. In what country(ies) did the music of this time flourish?

5. Who composed the music of this period?

6. Name some of the instruments used during this time period.

7. Listen to the music of this time period and describe what you hear. List some characteristics.

8. Listen to the music of this time period and explain what comes to your mind when you hear it.

9. Listen to the music of this time period and describe how it makes you feel when you hear it and explain what causes you to feel that way. For example, do you think it's the instruments, the tempo, the words or something else that you hear that causes you to feel the way that you do?

10. If you are in a classroom setting, listen to a particular song of this time period and create a scene **without words** that fits the music. Write it down then act out the scene. Have the class attempt to figure out the meaning of your scene. If you are studying music individually, listen to a particular song of this time period, create a story that fits the music. Record it in a journal.

Rap Music (1979-)

Historical Background: Towards the end of the 1970's, during the same time the media coined popular music that's dominating the charts as "pop", a new form of music emphasizing rhythm rather than melody or harmony developed. This music is called **Rap** and it was brought to the mainstream public by a group called the Sugar Hill Gang with its hit, *Rappers Delight.* The lyrics to *Rapper's Delight* was originally written by the pioneer MC, Grandmaster Caz of the group, The Cold Crush Brothers. Unfortunately, Grandmaster Caz never received the notoriety he deserved for being the originator of this ground-breaking music but his name, forever, will be synonymous with the original school of Hip Hop since he was the first person to mix the art of rhyming while D.Jing.

This innovative music was different because it consists of speaking (rather than singing) rhymes over rhythms and background music (minus the chorus and verses). Sometimes raps were comprised of just a steady rhythm with no melodies or harmonies that sometimes mirrored the funk and disco rhythms. With this music, no bands are used. Instead, synthesizers, drum machines and turntables accompany the rap artists.

About the Music: The importance of this music is not only the rhythm or sound it possesses but also like the blues of the 1920's and 1930's, and soul of the 1960's, rap music had a message. When rap music was first brought onto the scene, the message was light and humorous in that it pertained to partying or having fun. Some rap artists (male and female) would "battle" each other lyrically on the microphone in an attempt to reign as the "Master MC" or "Master Rapper." However, as rap music progressed, the messages it relayed became more serious in that it was used as a tool to voice the realities of urban life. With rappers such as Run DMC, the Furious Five, and Ice-T of the early to late 1980's and later with the rap group Public Enemy and also, Ice Cube (formerly of the Rap group N.W.A.) of the early 1990's, rap told the story of the social, economic and political injustices taking place in urban America. This brought about racial and political awareness, which in turn generated racial unity among African-Americans. Just like the European Nationalists of the Romantic Period during the nineteenth century (see Music During the *Romantic Period* 1827-1900), rap music brought about nationalism or a "racial oneness" within urban

America. It created a whole new culture within a culture because it became the voice of the African American youth living in the inner city.

As the voice of the African American youth continued to paint the picture of urban culture, another form of rap music with its use of strong language emerged. This hard-core form of rap is called **Gangster Rap** (pronounced "Gansta" Rap). Not only does gangster rap depict the social ills of growing up on the streets, also, it tells the story of and sometimes even glorifies sexually explicit, drug and gang related activities occurring in the inner cities of America. Rap artists such as the late Tupac Shakur and the late Nortorious BIG and others are known for this style of rap music. Unfortunately because of the violent nature associated with gangster rap, sometimes the poetic quality of the lyrics and the sincerity of some of the messages are missed. It is also shameful that because of its intense language, **all** of rap music, as a whole, has fallen prey to undeserved negative attention.

During the mid 1980's through the mid 1990's, while the messages of Rap were changing, the music went through changes as well. In addition to hard-hitting rhythms, rap artists began incorporating rhythmic percussion-like expressions with their mouths making them "human beat boxes." Artists such as Dougie Fresh of the Get Fresh Crew and Darren Robinson of the Fat Boys (formerly known as the Disco Three) were innovators in this area.

Hip Hop Music (1990-)

Historical Background/About the Music: Later as rap music continued to evolve, it became more melodic. It consisted of melodies that "accompanied" the rhythms and singing that coupled the speaking of the rhymes. This style of rap was pioneered by rap artists such as Dr. Dre and Snoop Dogg. With this change of style, **Hip Hop** was born. This music consists of hard-hitting rhythms, melodic instrumental parts (mainly synthesizers), "singable" melodies, and rap; a fusion of rhythm and blues and rap. A few other artists that dominated and continue to dominate this style of music are Queen Latifah, Mary J. Blige (known as the Queen of Hip Hop), Destiny's Child, Nelly and Jah Rule. But hip hop didn't stop there. Even though a technique called "sampling" (the mixing of earlier recordings with new music) had been used by artists such as the Sugar Hill Gang (mentioned earlier) and M.C. Hammer with his hit *Can't Touch This* (sampled from Rick James' *Super Freak*) in the early 1990's, Sean "Puffy" Combs also known as P. Diddy took it to another level when he began combining rhythm and blues and rock recordings from the 1980's with hip hop rhythms to create a contemporary sound. Clearly it is shown in his tribute to the slain Nortorious BIG with his song featuring Faith Evans, *I'll Be Missing You* (sampled from Sting's *Every Breath You Take*), which topped the charts in 1997. This technique became so popular that many artists such as Angie Stone with her song, *No More Rain* (sampled from Gladys Knight and the Pips' *Neither One of Us (Wants to Say Goodbye)*) and Mary J. Blige with her song, *No More Drama* (sampled from *Nadia's Theme* composed by Barry Devorzon and Perry Botkin and used in the opening of the soap opera, the *Young and the Restless*) took their songs to the top of the music charts using this method.

It is obvious that today's popular music would not exist if it weren't for the music that came before it. **Rap** and **hip hop** would not be what it is today if the sounds of gospel, rhythm and blues, rock and roll, soul, funk, disco and pop were never created. These earlier styles laid the foundation for today's popular music. Like any other style of music, rap and hip hop will change and develop into a new style. Ten years from now, the youth of today will look back on rap and hip hop and reminisce about [back in the day] when music reflected a time they knew as a youth; a time they lived as a youth. They'll remember when music reflected ***their own*** history."

Quiz 19

1. What are the approximate dates of this period?

2. What was the historical background of the music of this period? Explain the thoughts of the people or any major events taking place that had a direct effect on the music during this time period.

3. What type of music flourished during this time period?

4. In what country(ies) did the music of this time flourish?

5. Who composed the music of this period?

6. Name some of the instruments used during this time period.

7. Listen to the music of this time period and describe what you hear. List some characteristics.

8. Listen to the music of this time period and explain what comes to your mind when you hear it.

9. Listen to the music of this time period and describe how it makes you feel when you hear it and explain what causes you to feel that way. For example, do you think it's the instruments, the tempo, the words or something else that you hear that causes you to feel the way that you do?

10. If you are in a classroom setting, listen to a particular song of this time period and create a scene **without words** that fits the music. Write it down then act out the scene. Have the class attempt to figure out the meaning of your scene. If you are studying music individually, listen to a particular song of this time period, create a story that fits the music. Record it in a journal.

Music Timeline Exercise

Directions: Draw a music timeline from the Ragtime era through today's popular music. (Remember time periods may overlap.) Include important key musicians and the instruments used in each period. Then once the timeline is completed, research and write a report on your favorite period in music history from the Ragtime era through Hip Hop. Include important dates, and instruments. Give a historical background of some of the key musicians and/or composers and explain why they had such an impact on the period they dominated.

Ragtime ____ Hip Hop

Music Word Find

Z	B	O	O	G	B	S	P	I	R	U	N	P	A	R
Z	R	A	G	N	O	S	K	R	O	W	H	L	H	V
A	M	K	H	I	P	H	O	P	D	I	S	Y	W	S
J	A	Z	C	O	L	D	I	S	C	Z	T	X	I	P
Y	N	R	L	B	L	Z	R	A	G	H	M	D	N	W
R	R	E	K	O	S	N	Z	R	M	C	O	O	F	I
A	C	L	T	C	E	K	R	A	G	T	I	M	E	N
R	O	L	F	U	U	S	N	G	J	H	O	L	L	B
O	D	O	R	X	L	D	G	T	A	L	D	I	P	L
P	V	H	L	K	B	N	O	I	Z	X	O	S	H	O
M	Y	D	Y	L	I	K	V	C	D	N	C	O	H	P
E	X	L	U	W	Q	N	P	O	S	Z	R	A	C	V
T	V	E	S	L	A	U	T	I	R	I	P	S	R	I
N	S	I	U	R	S	F	B	D	C	R	D	X	H	V
O	Z	F	E	I	G	O	O	W	E	I	G	O	O	B
C	I	E	W	L	D	I	S	L	N	R	C	K	B	F
B	D	R	H	Y	T	H	L	A	C	K	R	V	O	N

Blues
Boogie Woogie
Bop
Contemporary Jazz
Cool Jazz
Disco
Field Holler
Funk
Ragtime
Rap
Rhythm and Blues
Spirituals
Swing
Work Song

Conclusion

All of the music discussed in this last section, from Ancient through twentieth century music, has a common bond. Each style of music has its own history, which reflects either events or ideas of the time in which it flourished. When times change, so does the music. When attitudes and beliefs change, music follows. Music mirrors the human condition and it tells the story of past and current events. Understanding its history leads to a deeper appreciation for it. Those who appreciate music will be able to relate to it and use it as a means of expression.

Author's Note: In an attempt to include a basic music history lesson of Western and American music into a music instruction book, some music and artists had to be omitted. It is not because the music or artists not discussed in this book are less important than those mentioned. It's because there is a wealth of music to explore and it's impossible to discuss **all** of the genres of music and their artists and maintain a basic comprehensible music history lesson.

BIBLIOGRAPHY

Berendt, Joachim E. *The Jazz Book: From Ragtime to Fusion.* (Revised by H and B. (The Jazz Book: From Ragtime to Fusion)Bredigkeit with Dan Morgenstern and Tim Nevill) Brooklyn: Lawrence Hill Books, 1989.

Blesh, Rudi and Janis Harriet. *They All Played Ragtime.* New York: Alfred A. Knopf, 1950.

Blumfield, Aaron. *The Art of Blues and Barrelhouse Piano Improvisation.* San Leonardo: P/F PUBL. CO., 1979.

Budds, Michael J. *Jazz in the Sixties.* Iowa City: University of Iowa Press, 1990.

Dance, Stanley. *The World of Duke Ellington.* New York: Charles Scribner's Sons, 1970.

Estell, Kenneth. *African America: Portrait of a People.* Detroit: Visible Ink Press, 1994.

Gridley, Mark C. *Jazz Styles: History and Analysis.* Englewood Cliffs: Prentice-Hall Inc., 1991.

Grout, Donald J. *A History of Western Music.* New York: W. W. Norton & Company, Inc. 1988.

Headington, Christopher. *History of Western Music.* New York: Schirmer Books, 1974.

Kenyon, Nicholas. *Authenticity and Early Music.* New York: Oxford University Press, 1991.

Lyons, Len. *The Great Jazz Pianists.* New York: Quill, 1983.

Machlin, Paul S. *Stride: The Music of Fats Waller.* Boston: Twayne Publishers, 1985.

Porter, Lewis and Michael Ullman. *Jazz: From Its Origins to the Present.* Englewood Cliffs: Prentice-Hall Inc., 1993.

Schafer, William J. and Riedel, Johannes. *The Art of Ragtime.* Louisiana: Louisiana University Press, 1973.

Schuller, Gunther. *Early Jazz.* New York: Oxford University Press, Inc., 1968.

Stewart, Rex. *Jazz Masters of the Thirties.* New York: The Macmillan Company, 1972.

Tate, Eleanora E. African American Musicians. New York: John Wiley and Sons, Inc., 2000.

Taylor, Billy. *Jazz Piano.* Dubuque: W.C. Brown Company Publishers, 1982.

Weiss, Piero and Taruskin, Richard. *Music in the Western World: A History in Documents*. New York: Schirmer Books, 1984.

About the Author

Dr. Darshell DuBose-Stone was introduced to music at an early age and was trained as a classical pianist throughout her teenage years. When she was accepted at Lycoming College in Williamsport, Pennsylvania she began to study Ragtime and Jazz piano and their history. Therefore, in addition to studying the required great European composers and their music, DuBose-Stone devoted much time to an independent study on the progression of jazz piano for which she received a Departmental Honor and was inducted into Phi Alpha Theta, an International Honor History Society. This was the beginning of her journey as a music historian.

After graduating from Lycoming College with honors in 1992, DuBose-Stone decided that she wanted to continue her studies in Music History and was accepted into the master's program at Rutgers University. It was during this time that she studied at the Jazz Institute and devised her master's thesis, *Learning How to Play Jazz Piano from the Beginner's Point of View.*

Shortly after graduating from Rutgers University in 1994, DuBose-Stone began teaching piano performance at several reputable music studios in New Jersey until she decided to build her own private and online music studio, Darshell's Quality Music which is still active today. During this time, she devised an innovative method of teaching and put it into a curriculum which she uses on her own students. The curriculum produces favorable results. Her students comprehend the fundamentals of music with little or no difficulty. Also, they learn to play the piano, drums and/or sing and become educated in music theory, history, and analysis. Therefore, in 1998, she made the decision to expand her teaching services to reach outside of the four walls of her music studio and developed music curricula for various institutions such as the largest state juvenile secure-care facility in New Jersey, a Freedom School program, and a small New Jersey independent charter school.

Out of this curriculum, D.S. Education Alliance, LLC, an instructional company was established by DuBose-Stone and her late husband, Jerel Stone. This company provides education in various disciplines to schools, churches, secure-care facilities, camps, and private households. Under the umbrella of D.S. Education Alliance, the Freedom Tour was created. This program gives music and dance students (trained by various educators) the opportunity to travel to historic cities throughout the United States and showcase their talents. In addition, under D.S. Education Alliance, LLC, DuBose-Stone partners with educators of various disciplines who provide instructional workshops that meet the learning needs of students in all types of educational institutions. Upon completing her doctoral studies in curriculum and education at Northcentral University/National University, DuBose-Stone continues to serve as a certified music educator and writes music curricula for educational institutions that request well-rounded music programs to be established. She has been recognized by the National Association of Professional Women (NAPW), the National Society of Leadership and Success (NSLS), and the International Honor Society of Kappa Delta Pi.

For more information on D.S. Education Alliance, LLC, visit www.dseducationalliance.com or contact us at (201) 370-8664 or darshell@dseducationalliance.com.

Special thanks to Marc Medley for editing the puzzles.

Made in United States
Orlando, FL
04 November 2026

90207834R00087